MALAYSIA AIRLINES FLIGHT 370

MALAYSIA AIRLINES
FLIGHT 370

WHY IT DISAPPEARED—AND WHY IT'S ONLY A
MATTER OF TIME BEFORE THIS HAPPENS AGAIN

DAVID SOUCIE

Skyhorse Publishing

Skyhorse Publishing books may be purchased in bulk at special discounts for sales promotion, corporate gifts, fund-raising, or educational purposes. Special editions can also be created to specifications. For details, contact the Special Sales Department, Skyhorse Publishing, 307 West 36th Street, 11th Floor, New York, NY 10018 or info@skyhorsepublishing.com.

Skyhorse® and Skyhorse Publishing® are registered trademarks of Skyhorse Publishing, Inc.®, a Delaware corporation.

Visit our website at www.skyhorsepublishing.com.

10 9 8 7 6 5 4 3 2 1

Library of Congress Cataloging-in-Publication Data is available on file.

Cover design by Brian Peterson
Cover photo: Associated Press

ISBN: 978-1-63220-729-6
Ebook ISBN: 978-1-63220-849-1
Printed in the United States of America

CONTENTS

PREFACE

The disappearance of Malaysia Airlines Flight MH370, with 239 people on board, reveals profound vulnerabilities in air travel. The tragedy has dramatically, horrifically changed the lives of the families and friends of the victims and has left the rest of the world in shock, stunned that a giant plane seems to have vanished—without a trace—and no one has answered why. To most people, an event such as this was unfathomable. To those of us who have investigated aircraft accidents, it's a given that some planes won't make it to their intended destinations. When we investigate, what we need to find out is why. Why didn't the plane make it?

Countless articles and blogs have speculated about what might have happened to MH370. Many of them use the disappearance of MH370 as a muse for wild speculation and the development—and spewing—of fictional theories. This book is different. I will provide you with a concise, reasoned, and rational approach to understand what most likely happened to the aircraft. I'm not trying to convince you of anything—I have no agenda other than to allow you to have all the facts to make an informed judgment call. The evidence speaks for itself.

You will notice that this book does not address the ongoing search in great detail. I am not a satellite specialist and do not have the expertise to judge the validity of the information nor the calculations. I have therefore elected not to address how the technology works or how the scientists involved in the search arrived at their conclusions. There are independent groups who have also analyzed the data that was released from the investigation team. There is plenty of information available on the Internet about this technology and where the search is focused. This book will not provide you with that information.

The call to write this book came from Skyhorse Publishing while I was in Los Angeles filming a documentary on the disappearance of MH370 for the History Channel. Skyhorse published my first book, *Why Planes Crash: An Accident Investigator's Fight for Safe Skies*, reissued under the title *Safer Skies*, and without hesitation I jumped at the chance to work with Skyhorse again. I did, however, have some reservations about writing a book on this particular topic: What happened to MH370? I worried about coming across as disrespectful to the families of the missing passengers and crew. What eased this fear is what I've seen over and over during my many years as an accident investigator. It is the common thread that all accidents share: The families and friends of the victims have a need to understand what caused the accident. There's also an unyielding desire to assure that an accident of that type never happens again.

To figure out what most likely happened to Flight 370, I designed a type of Bayesian analysis model. Bayesian analysis—or Bayesian probability—is named after the great eighteenth-century mathematician Thomas Bayes and is a logic-based system used to assess the likelihood that an event occurred. The model I used to determine what most likely happened to MH370 is a fairly simple mathematical formula. It weighed each piece of information about the flight against others. It considered the reliability of sources and evaluated information on its own merit. The method, developed through my many years of experience as an aviation safety inspector and analyst, took into account what I know about quantifying the importance of circumstantial and physical evidence—clues. It assigned values to things like possible human motives—for example, did a passenger or crew member sabotage the flight? It also assigned values to hard information, such as what events, both common and unexpected, could have caused a mechanical malfunction.

There were times during the process of writing this book when I wondered if I'd gotten myself in over my head. What kept me going was correspondence I had with Sarah Bajc. Sarah's partner, Phillip Woods, was on board MH370, and she has gallantly taken on the role of spokesperson for the families of the MH370 passengers. She's also been the catalyst motivating several independent groups working to make sure the investigation stays the course. Dr. David Gallo, Director of Special Projects at the Woods Hole Oceanographic Institution, introduced me to Sarah through email. In her first email to me, Sarah called me "bright and trustworthy"—something she said was in short supply during the investigation—and she asked me to continue trying to solve the mystery of MH370. Without Sarah's support, this book might not have come to fruition.

It's now been more than a year since MH370 disappeared, yet interest in the mystery of Flight 370 hasn't let up. There are the pundits who continually spout their claims about what could have happened, and then there are people like me who make informed assessments. I hope I'm one of the people you actually want to hear from. I'm aware of what leads to real answers and why wild, unsubstantiated claims don't help anyone. The world deserves to know what happened to MH370, the families of the victims need closure, and the aviation industry needs to address its shortcomings.

This book is my best effort toward making that happen.

1

EARLY MISTAKES

It was 1:19 a.m. in Malaysia on March 8, 2014, when the captain of Malaysia Airlines Flight 370, Zaharie Ahmad Shah, said good night to the air traffic controller in Ho Chi Minh City. It would be the last communication from the ill-fated flight. Two minutes later, air traffic controllers in Kuala Lumpur, Malaysia, watched the aircraft on their radar screens as it passed over its planned waypoint, a place called IGARI. Within ten seconds of passing IGARI, the radar label for Malaysia Airlines MH370 disappeared from the radar screen. Its transponder had been deactivated.

There were no distress calls from the aircraft, no indication anything was wrong, yet five independent communication systems on board the Boeing 777 aircraft went silent. It was just eighty minutes after the flight took off from Kuala Lumpur International Airport and the aircraft, its 239 passengers and crew members had vanished.

Air traffic controllers frantically searched for an answer, but they were unprepared for anything like this, and as a result were unable to effectively respond to it. Grave errors were made, setting the stage for what may be a permanent veil of mystery over what happened to MH370. The errors ranged from blundering ignorance to missteps to blatant arrogance. The inevitable talk of suspected cover-ups would ultimately result in a shuffle of Malaysian leadership, further compounding the complexity of the search. The first few of these blunders were the most costly of all. Had they not occurred, we would likely know exactly where the aircraft is. If there were survivors, they might have been rescued.

The first mistake occurred when MH370's flight trackers mistakenly reported that the plane was flying on its scheduled route over Vietnam when, in fact, the aircraft had made a dramatic change in direction away from Vietnam, heading south. This misinformation delayed for several hours the start of search and rescue (SAR) operations. It also delayed reports of the plane's disappearance to the Malaysian authorities, wasting even more precious hours. Together, a full eight hours were wasted by these delays.

Another massive mistake was that Malaysian civilian radar operators, Malaysia's air traffic control, did

not immediately alert Malaysian military radar operators when they realized that the plane had gone missing. A Malaysian military radar operator was actively
tracking a radar image of Flight 370 after it went missing, but the operator had no idea that a search for it
was underway. No one had told him. It would be days
before the military radar operators even heard of the
investigation. If there had been early communication
between civilian and military radar operators, the
significance of the sighting could have been pursued
right away.

While time was wasted, the families of the missing
passengers and crew members agonized, not knowing if their loved ones were alive. During this time,
Sarah Bajc, partner of missing passenger Phillip Wood,
insisted there were survivors waiting for rescue. If there
were survivors, were they lost at sea or had the plane
landed somewhere unexpected? Eventually, the families of the missing passengers and crew members began
losing hope as the search changed from a search and
rescue operation into a search and recover one.

When people ask me, "Why should this search continue?" or "Why are we spending millions of dollars
trying to find the plane?" my answer is, "We must."
We need to know if an individual was responsible for
all of this or if there are larger systemic failures in the
aviation industry. We need to feel safe when we travel
from country to country, and we need to know that air

traffic controllers are paying attention to our flight—that they haven't drifted off to sleep and lost the signal from our airplane. And most importantly, we need to know that the lives of those on board MH370 were not lost in vain and put their memories to rest.

The search for Flight 370 and the 239 people on board has recalibrated my view of space and time. On the one hand, the ability to travel across oceans in mere hours has made the world seem smaller. On the other, MH370 has provided a harsh dose of reality: it shows just how small we really are and how unforgiving this planet can be. There's a whole world out there that we could get lost in.

2

SILENT RUNNING

Captain Zaharie Ahmad Shah placed his bags on the X-ray machine and walked through the security scanner in Malaysia's Kuala Lumpur International Airport around 11:00 p.m. Malaysian time (MYT) on March 7, 2014. It was a familiar routine for him; he had worked for Malaysia Airlines for more than thirty years. That night, he would be flying a Boeing 777 to Beijing. His first officer, twenty-seven-year-old Fariq Abdul Hamid was new to the Boeing 777 but had years of experience flying smaller aircraft. Fariq also placed his bag on the X-ray machine and passed through the security scanner. At 12:32, the aircraft was cleared by Kuala Lumpur ground control to taxi to holding point Alpha 11, Runway 32, right. At 12:41, the plane took off. It was passed from communications with ground air traffic control to outbound radar control. At 12:46, it was cleared to fly directly to IGARI. At 12:50, it was cleared to climb to 35,000 feet.

This chain of events is routine, not at all unusual. As a former Federal Aviation Administration (FAA) safety inspector, I have seventeen years' experience observing and evaluating cockpit crews from the jump seat. I can recognize when pilots are hiding something—if they're not 100 percent sure of their skills, for example—and when they're very comfortable with their abilities. As I listened to the voice recordings of Zaharie and Fariq as they proceeded through the rote responses to the controllers, I recognized calm and confident composures. They gave the responses expected from pilots of their caliber.

For the next eleven minutes, the aircraft climbed to 35,000 feet. At 1:01, Captain Zaharie reported to Kuala Lumpur radar control, "Maintaining flight level three five zero."

While rising, the aircraft made a scheduled call from the Aircraft Communications Addressing and Reporting System (ACARS), a messaging system, to a geosynchronous satellite far overhead. By sending and receiving a series of digits, the satellite identified the call as being from MH370 and answered it. Data communication from an aircraft is protected from hacking by a twenty-four bit hexadecimal encoding system, and once the connection was established, MH370 sent information to the satellite, which was relayed to an information management services facility on the ground. From there, the information was passed on to the Malaysia Airlines communication center. The entire process, which is standard, took only seconds to finish.

ACARS is the most commonly used system for aircraft and pilots to communicate with air traffic control. It works as an air-to-ground link for data and voice services and was initially designed and deployed in 1978 by Aeronautical Radio, Incorporated (ARINC), a US company with regional headquarters around the world. The early versions of ACARS used only air-radio communications; later, ARINC teamed up with Inmarsat, a London-based company, to provide mid-continent satellite communications and with Iridium, a US corporation that provides global voice and data satellite communications in the southern- and northern-pole regions.

Today, ACARS is considered a critical part of the safety umbrella for air and maritime operations. It works in a method similar to that of a smart phone with a voice and data package. Messages are automatically sent from the aircraft every half hour, and they include a great deal of information about the aircraft in data packets. Depending on the level of service an airline purchases, the packets can include information about everything from the health of an aircraft's engines to the movement of its flight controls. In the case of MH370, key information on the plane's location, speed, direction, and mechanical condition was supposed to be relayed to a satellite every thirty minutes; however, the second to last data transmission passed from MH370 through ACARS at 1:07, and fifty-five seconds later, Captain Zaharie keyed the microphone for the second to last voice message.

01:07:55: "Malaysian Three Seven Zero maintaining level three five zero."

An air traffic controller in the Kuala Lumpur radar center in Subang, Malaysia, acknowledged the call at 1:08, registering the message with a simple, "Malaysian Three Seven Zero."

For eleven minutes and twenty-four seconds, there was no communication between air traffic control and MH370. Captain Shah and First Officer Hamid continued flying straight toward Beijing and level at 35,000 feet.

At 01:19:24, in another routine exchange, Kuala Lumpur air traffic control contacted MH370's pilots to let them know they were transferring control of the aircraft to the next radar center in Ho Chi Minh City. ATC said, "Malaysian Three Seven Zero contact Ho Chi Minh 120 decimal nine. Good night."

Captain Shah responded cordially at 01:19:29, "Good night, Malaysian Three Seven Zero."

After the call, neither Captain Shah nor First Officer Hamid contacted Ho Chi Minh City—and the aircraft made no automated transmissions to it.

When the radar controller ended communication, he officially completed his job. He'd guided MH370 to the IGARI waypoint and provided the pilot with the information he needed to contact the next radar control center along his path to Beijing. It would have been customary, though not required, for the controller to then radio the Ho Chi Minh radar center to let them know MH370 would be contacting them and entering its controlled airspace. I have not been able to confirm if that happened or not. I suspect it did because the Ho Chi Minh radar center contacted the Kuala Lumpur center seventeen minutes later, at 1:38, and asked where the aircraft was, so they were clearly expecting it—and wondering why it hadn't made contact.

Prior to the call from the Ho Chi Minh ATC, the Kuala Lumpur radar controller either didn't notice or didn't report that the aircraft's information had disappeared from his screen after his call with the MH370 pilots. The disappearance of the information means that MH370's transponder was turned off. Why is this important? In order to answer that, one needs to understand how aircraft radar and transponders work.

Aircraft radar has two components: primary surveillance radar (PSR) and secondary surveillance radar (SSR). Primary surveillance radar measures the time it takes for electromagnetic waves to travel from an antenna to an aircraft and return. This shows the aircraft's distance from the antenna and its bearing, or the direction it's traveling in. The bearing and distance position the plane on an ATC display screen so its controllers can track it.

The advantage of primary surveillance radar is that it requires no action from an aircraft in order to provide a radar return. The radar bounces off an aircraft's surface and circles back to the antenna. Disadvantages of PSR are the enormous amount of power sent from the antenna and the small amount of signal returned from the aircraft. The small return signal makes it weak and allows it to be easily interrupted by things like rain. The enormous power of the antenna, intended to compensate for this, makes it extremely sensitive to motion, and it often picks up distracting, meaningless ghost images from ground clutter.

PSR does not automatically identify an aircraft. In the early days of its usage in commercial aircraft travel, a controller would identify a plane by requesting its pilots turn it left or right, and when a blip on the controller's screen changed course in the requested direction, it was identified. Of course, this method was inefficient, and when the skies began to fill with commercial jets in the 1960s, another technology had to be devised. This is when secondary surveillance radar was born. Like PSR, SSR measures an aircraft's range and bearing. It also provides a plane's altitude, speed, and flight number to ATC.

SSR works by requesting the information it needs from a piece of equipment on board an aircraft known as a transponder, which consists of a radio receiver and transmitter. Today, information from SSR is relied on by civilian radar much more than that of PSR. The range of either radar is typically between 200 and 250 miles. In the case of MH370, the SSR information stopped showing up on the ATC screen because the transponder stopped responding to the requests. Military radar, which tracks PSR, could have seen the plane—and, as I mentioned, one military radar operator did see it; he just didn't realize at the time that it was noteworthy. He didn't know to report it until later.

At a news conference on March 12, 2014, Datuk Azharuddin Abdul Rahman, Director General of the Malaysian government's Department of Civil Aviation, stated that MH370's transponder stopped communicating at 1:21 a.m. Until that time, everything about MH370's flight path and communication seemed routine. After 1:21 a.m., nothing about these seemed routine. When Kuala Lumpur ATC realized it had lost contact with MH370 and the plane hadn't made contact with Ho Chi Minh ATC, they knew something had gone horribly wrong. Repeated attempts to contact the aircraft were made by both Kuala Lumpur and Ho Chi Minh ATC—calls were made on every frequency—yet there was no response. Ho Chi Minh ATC had seen a radar image of MH370 flying over waypoint BITOD, which is thirty-seven nautical miles past the IGARI waypoint on what should have been the plane's path to Beijing, but the image was lost soon after that and couldn't be reestablished. Other aircraft in the vicinity of where MH370 was last known to be were made aware of the missing aircraft and also attempted to contact it, but their attempts at communication also failed. No one could reach the plane nor its pilots.

3

PLACING BLAME

The search and rescue (SAR) operation for MH370 began in the Gulf of Thailand and the South China Sea four hours after the Ho Chi Minh air traffic control reported that the MH370 pilot had failed to contact them. The Malaysian government and Malaysia Airlines were severely criticized by the international media and social media for the four-hour response time. Particularly critical, of course, were the families of the passengers and crew. The possibility of finding survivors seemed less and less viable every minute the search was delayed.

Was the criticism warranted? Dispatching search and rescue personnel and equipment too early, without reliable information and a well-thought-out plan for where to search, is futile in situations such as the search for MH370 and can hamper the operation. Sending search teams and equipment in the wrong direction and then having to recall them due to new, more accurate information wastes valuable time. The management and proper placement of resources can make the difference between life and death for survivors. Because there was little to no dependable information on the whereabouts of MH370, the early decision to dispatch SAR resources into the Gulf of Thailand and the South China Sea was a waste.

As a professional safety investigator, I prefer to not retrospectively criticize accident investigations. Dispatching the resources was a waste, but I don't blame the Malaysian authorities for doing it—they had to do something. But their resources would have been better spent investigating more radar facilities and communicating with adjacent countries. While they searched the Gulf of Thailand and the South China Sea, MH370 was still in the air, thousands of miles south of where they were searching for it. In fact, Thai military radar was watching MH370 cross its radar screen, but because the unidentified aircraft was not a clear and present danger to the coast of Thailand, the radar operator disregarded it. It was logged as a commercial jet that posed no threat. The Thai military radar operator, like the Malaysian military one, didn't know early enough that what he saw was noteworthy.

The United States and several other countries have chosen to integrate military and civilian radar information. Prior to the attacks of 9/11, this was not the case, and this made it impossible for the military to identify and mitigate the threat from a hijacked commercial aircraft. Today, in many parts of the world military and civilian air traffic controllers sit side-by-side in control centers. This is not the case in Malaysia.

The possibility that there were people alive on board the aircraft and a better search and rescue operation could have saved them is almost too troubling to consider. It brings to light what may be the greatest vulnerability in air travel, which is the holes in international communication. If the plane is found and a comprehensive analysis of the search reveals the passengers and crew could have been rescued, it might implicate the failures of the international safety system as a contributing factor to the loss of life. Perhaps if the system better connected its member states, the search could have been more effective in its early stages and lives could have been saved.

Considering this possibility raises a number of questions. To begin with, how is it possible that after billions of dollars were spent by airlines and governments and member states of the International Civil Aviation Organization (ICAO)—a United Nations agency that develops international standards and recommended practices (SARPs) for civil aviation requirements— that this could happen? Who is accountable for the holes in the system? And who will fix it?

If the aircraft is found and a final report is issued by the Malaysian Department of Civil Aviation (DCA), a perfect blame storm will follow. Finger pointing and litigation by the families of the passengers and crew will make up the first wave of attacks, then civil aviation authorities will begin their own finger pointing. Airlines will blame ICAO, and ICAO will blame the International Air Transport Association (IATA), which is the trade association representing some 240 airlines. The Australian Transportation Safety Bureau (ATSB) and Malaysian DCA will make recommendations for improvements in civil aviation guidelines, including international regulatory reforms by ICAO. There will be accusations by various parties that the recommendations put forth in 2012 by France's Bureau d'Enquêtes et d'Analyses pour la sécurité de l'aviation civile (BEA), its official aviation accident investigators, weren't followed and clearly should have been. Almost all of these accusations will be difficult to dodge.

4

GROUNDHOG DAY

Often during my investigations, I feel like I'm having a *Groundhog Day* experience. It's as if I'm Bill Murray's character, and I'm reliving the same day over and over again. Wake up? Check. Have breakfast? Check. Hear about a plane that disappeared or crashed but might not have if there were better safety measures in place? Check.

The day I heard that MH370 had vanished, I felt this sweeping over me again. It seemed like I was watching something I'd already seen unfold, and then it hit me how much MH370 reminded me of Air France Flight 447, an Airbus A330 that went missing in 2009. AF447 was traveling from Rio de Janeiro to Paris and was supposed to land there on June 1, 2009. Needless to say, it never made it. It suffered from a high altitude stall and crashed into the Atlantic Ocean. The plane was carrying 228 people, and all were killed in the crash. When the aircraft was finally located two years after it went missing, BEA launched a full investigation into what happened to it and why and wrote up an extensive list of safety recommendations for commercial aircraft to use going forward. Perhaps if they were all in practice now, MH370 might not have gone missing or at the very least would have already been found. Why does it seem like this continually happens, yet no real preventative measures are taken?

What allowed investigators to find AF447 is something that the investigators of MH370 don't have: ACARS transmissions sent to Inmarsat satellites every ten minutes until just before the plane crashed. AF447's last ACARS transmission was sent at 02:10:34 a.m. UTC time on June 1, 2009. It gave GPS coordinates of 2.98° N latitude/30.59° W longitude, a ground speed of 107 knots, and notice that the aircraft was descending at 10,912 feet per minute. This information allowed investigators to narrow down the area where the plane most likely was, and using oceanic drift models and complex mathematics, they were able to launch their search.

For the search, BEA contracted two tugs to scour the ocean floor looking for the two recording devices— the cockpit voice recorder (CVR) and the flight data recorder (FDR)—together known as the plane's black box, which would reveal what happened to the plane. The tugs contracted for the search were the *Fairmount Glacier* and the *Fairmount Expedition*. Each tug was fitted with a towed pinger locator (TPL) system, which is capable of listening for a 37.5 kHz frequency transmission—the frequency transmission made from an underwater locator beacon (ULB). ULBs are mounted to a black box and activated to transmit a pulse or ping that can be picked up by a towed pinger locator hydrophone, or towfish, when it comes into contact with water. In the case of AF447, like with MH370, the ULB had a battery life of thirty days.

Search and rescue operations were conducted by the French and Brazilian authorities, and two towfish were borrowed from the United States Navy to help with the search. The towfish were attached to the end of a tow cable, which was released and retracted by a large winch on the tugs. By adjusting the speed of the tugs and the length of the cables, the towfish moved through water just above the bottom of the ocean. In the area of the Atlantic Ocean where AF447 crashed, it can get as deep as 14,000 feet below the surface. In order to scour the ocean floor at such depths, the towfish cable had to extend nearly six miles from the ships that towed them.

Within five days of the accident, the Brazilian Navy found and removed wreckage and two bodies from the ocean, but no pings were picked up by the towfish within the thirty days that its battery presumably lasted. Without being able to pick up the pings, investigators were left to comb the area without knowing exactly where to look for the black boxes. The black boxes were eventually found in April 2011, and finally the cause of the crash, a high altitude stall, was officially determined.

After the aircraft was found, France's BEA performed hundreds of safety tests and made many safety recommendations to the European Aviation Safety Agency (EASA) and the International Civil Aviation Organization (ICAO), one of which was to "extend as rapidly as possible to 90 days the regulatory transmission time for ULBs installed on flight recorders on airplanes performing public transport flights over maritime areas."

The FAA supported this recommendation and, via a letter dated January 28, 2010, requested that SAE International, an association that sets technical standards in industries including aerospace, form an industry working group to call for an increase in the minimum operating life of ULBs from thirty days to ninety days. SAE International did so, but in classic FAA form, the rule they decided to implement was one that allowed for an as-minimal-as-possible financial toll on the airlines, requiring airlines to replace their thirty-day units with ninety-day units only when the batteries are retired from service. The result of this is that thousands of aircraft, including MH370, still have the thirty-day units even though BEA presented this recommendation back in 2009. Given this, it might not be long before another plane's black boxes need to be found and the world again watches thirty days come and go knowing that having ninety days may have allowed an aircraft to be found sooner than in two years, as in the case of AF447, or however many years it takes for investigators to find MH370, if it is ever found. Those extra days could save millions of dollars and months or years of anxious waiting for closure by the families and friends of the victims of a crash.

5

THE COST OF NOT FINDING MH370

The search for MH370 is proving to be the most costly ever made for an aircraft. In November 2014, the Malaysian government revealed the combined amount it had spent on the search and rescue operations for MH370 and MH17, the Malaysia Airlines flight that was shot down over Ukraine in July 2014. The amount was a staggering 33,461,861.50 Malaysian ringgit, which is about 10 million dollars. By far, most of the funds went to the search for MH370, and of course, the search is ongoing. The Australian government has committed a whopping 80 million dollars for a two-year search. And the search has also taken a large financial toll on China, where most of MH370's passengers were from, and others who've pitched in to help, such as the United States, the United Kingdom, New Zealand,

Japan, and South Korea. There is, however, a limit to how much the world will deplete its pocketbook. If there are no tangible results, the investigators will at some point need to recognize defeat and abandon the search. Their governments can't afford for them not to.

I know from experience how the lack of tangible evidence gnaws at an accident investigator and how eventually the passion driving gallant efforts to find answers is replaced with feelings of futility and hopelessness. The days, weeks, and months spent searching for a missing aircraft are eventually spent justifying the money that was spent. The real question, however, is not how much it costs to find the plane, it's how much it will cost if it's *not* found. How much will it cost if investigators don't determine and can't publicly declare the cause of MH370's disappearance? And what if they can't assure that what happened to MH370 won't happen again?

If the MH370 aircraft is not found and measures to mitigate the tragedy are not taken, the consequences will be disastrous. Potential passengers will, quite reasonably, wonder if whatever it is that happened to MH370 could happen to a plane that they board. This is a massive problem in many, many ways. Financially speaking, millions of dollars are being lost by Malaysia Airlines as its jets fly through the skies with vacant seats. In August 2014, Malaysia Airlines was removed from the Malaysian stock exchange as part of a twelve-point recovery plan in an effort to rebound from its losses. The company is in the process of being privatized. It reported that its net loss in the July–September quarter of 2014 rose 53 percent from a year earlier to approximately 170.3 million dollars, and their loss is rapidly increasing.

Beyond its effect on Malaysia Airlines, if the mystery of MH370 is not adequately resolved, people will be left wondering if all commercial airlines lack the requisite safety measures to care for their passengers and if the governments that oversee them are actually equipped to do so. People who planned to travel the world—to get to know other cultures—may cancel their plans or avoid making them altogether. Companies might lose money because their employees refuse to travel for business. The world that's been opened up to us by the incredible developments by the likes of Sir George Cayley, Otto Lilienthal, and the Wright brothers and was pioneered by the likes of Amelia Earhart—who herself went missing—will suffer.

Simply put: passengers will not fly if they don't feel safe, and only by taking the necessary measures to avoid planes from disappearing again can confidence in air travel be restored. Without question, the search for MH370 must be continued at all costs. I pray that the importance of this is internationally recognized and that the commitments that have been made to find the plane are held and realized.

6

COST-BENEFIT ANALYSES

In order to evaluate a decision to invest in a safety improvement, two factors must be analyzed: the cost and the benefit. Seems simple, doesn't it? The cost factor is determined by the losses that could result from not investing in the safety improvement. The benefit factor is less clear.

Here's an example: Suppose you have been appointed CEO of a major airline, and after BEA released its safety recommendations, your safety manager provided you with reams of information about the thirty-day life of the ULB batteries mounted to the cockpit voice recorder and the flight data recorder. After reading the documents, you are convinced that ULBs should transmit pings for more than thirty days and want your company's devices to do so. You speak with your employees about this and are informed that the cost of upgrading the units to a recommended life of ninety days is 3,000 dollars per plane (2,000 dollars for the batteries and 1,000 dollars for the labor), so the cost for upgrading your 700 aircraft is 2,100,000 dollars. That is a lot of money, you think. So is it really worth it?

The benefit of spending the 2,100,000 dollars to upgrade the ULBs is an improved ability to find the black boxes if one of your planes ends up in the middle of an ocean. But as your comptroller will surely remind you, this rarely happens. In fact, he does remind you of this when you bring it up in your Monday morning staff meeting. He looks at you, well-dressed and teeth sparkling, and tells you that 2,100,000 dollars could fund seven more Sydney-LAX trips each week, that those trips are currently earning about 900,000 dollars in revenue each, and that by adding this revenue to the current number of Sydney-LAX trips your airline has, it won't take long to earn out the 2,100,000 dollars and then net profits will rise rapidly.

"I am sure that ULB thing is a very important safety investment," he says, "but imagine how much money we can make if we invest the two million differently. Down the road, we can reconsider this."

Not so easy now, is it? Think about it the other way around though. What will it cost if you don't invest in safety? How much money will be spent if you have to search for the black boxes? How much money will be lost if ticket sales decrease? And even more importantly, what would it cost you personally if you have to look into the faces of family members of deceased passengers or families of the crew employed by your company knowing you didn't do everything you could do to find the missing plane their loved ones were on?

Jay Pardee, former manager of the Engine and Propeller Directorate of the FAA, the unit of the FAA responsible for certifications to new or changed designs of aircraft engines and propellers and one of my mentors, would refer to this method of thinking as shifting the paradigm from looking backward and diagnosing the problem to looking forward to costs before an accident happens. The cost of something that might happen is difficult to predict. The benefit, too, is difficult because it's only realized if something terrible doesn't happen.

Thinking beyond ATC batteries—or outside the box, so to speak—there can never be a guarantee that any one safety investment in an aircraft will prevent a loss of life. If a plane is in the air and there are passengers on it, there is always and will always be risk of this. It's the same as riding in a car. So questions of cost and benefit will come up every time the possibility of added measures to protect planes is discussed by airline owners and their finance managers.

Throughout the world, safety managers need to stay ready for these conversations and be willing to plod into accountants' offices and grand boardrooms to battle for safety. The financial cost of the loss of MH370 is certainly one thing they can reference if the emotional tolls and a company's reputation are not enough.

Other safety improvements recommended in the BEA report include transmission of aircraft information in real time to satellites and thus ATC stations and other methods of signaling the location of a downed aircraft. These improvements, had they been implemented when they were originally proposed or anytime up until March 8, 2014, might have prevented or dramatically changed the outcome of the disappearance of Flight 370.

On May 12, 2014, the International Civil Aviation Organization hosted a conference on aircraft tracking. Rupert Pearce, CEO of Inmarsat, said the company could provide basic free service tracking immediately over its existing L-band satellite network. He went on to explain that other fee-based services could also be taken to assure another aircraft doesn't simply vanish.

Pearce recommended satellite contact with planes every fifteen minutes rather than every hour or half hour; in such a plan, an aircraft would send positional GPS data, such as speed, direction, and altitude. There are also many other items he described that could be implemented without adding equipment to the aircraft, such as black-box-in-the-sky reporting, which I won't get into now.

All the recommendations Pearce gave are recommendations made to the ICAO and other civil aviation authorities in the report on AF447 four years earlier. Why then has not one of them been made mandatory by these same civil authorities?

I don't want to seem overly negative. There have been great safety programs designed to upgrade civil aviation. In the United States, for example, there is the Next Generation Air Transportation System (NextGen). NextGen was developed by the FAA and, among other things, intends to update the methods used for the satellite tracking of planes by installing the most modern forms of GPS technology and strengthening data communication between air traffic control and aircraft. The European equivalent to NextGen is the Single European Sky Air Traffic Management Research Program (SESAR). The problem is that these programs have been discussed for years now, yet their implementation is continually hampered by budget shortfalls. I hope that they will soon be fully funded and that this will cause a swift implementation of them, but judging from the aviation industry's history, I doubt this will happen.

In 1997, Mary Schiavo, former Transportation Department Inspector General, spoke in front of Congress following the ValuJet Flight 592 crash and referred to the US Safety Agency as a "tombstone agency." Why? "Because," she said, "they wait for

major loss of life before they make a safety change."
I wish I could say that things have changed since
1997 and the issues that concerned Schiavo have been
addressed, but it's far from that. I am convinced that
even tombstones would not be sufficient to motivate
civil aviation authorities worldwide to require contin-
uous in-flight tracking. Since there already are tombstones, it seems I've been proven right.

7

THE TEXT MESSAGE

The announcement came later than was scheduled. It was nearly two hours after the planned time on March 24, 2014, when Malaysia's Prime Minister Najib Razac arrived in a pitch black suit and sauntered up to the podium. As he began speaking, it was immediately evident that he was carrying the burden of bad news. Inmarsat had completed a study of data from a satellite that had received the final known signals from MH370 as it moved south. Razac explained:

> Using a type of analysis never before used in an investigation of this sort . . . Inmarsat and the AAIB [the UK Department for Transport's Air Accidents Investigation Branch] have concluded that MH370 flew along the southern corridor, and that its last position was in the middle of the Indian Ocean, west of

Perth. This is a remote location, far from any possible landing sites. It is therefore with deep sadness and regret that I must inform you that, according to this new data, Flight MH370 ended in the southern Indian Ocean.

I was watching the announcement across the world in the New York City green room at CNN as I prepared my analysis for yet another interview. I felt chills go down my spine as I heard the agonizing screams and sorrowful sobs echoing through the room in which the Prime Minister was speaking. The screams were all too familiar to me. They brought back memories of times I had to deliver similarly devastating news to families of other aircraft accident victims. As I took in what the Prime Minister had said, those memories that I thought had been long stored away came to the forefront of my mind as if the experiences had happened only days before.

In the wake of the announcement, an elderly man—presumably a family member of one of the victims—collapsed and was carried out of the conference on a stretcher. Moments later, a woman was taken out on another stretcher. The camera caught her staring, expressionless, seemingly in shock from the news.

An hour before the conference, a number of the victims' family members had the news broken to them through a text message from Malaysia Airlines.

The text was in English even though this was not the native language of most recipients. It read: "Malaysia Airlines deeply regrets that we have to assume beyond any reasonable doubt that MH370 has been lost and that none of those on board survived."

8

OCEAN SHIELD

On April 1, 2014, *Ocean Shield*, a massive red-hulled ship belonging to the Australian Navy, set out from Garden Island, Australia, on a voyage to a remote area of the Indian Ocean where, based on calculations by Inmarsat, investigators hoped to find evidence of the MH370 aircraft. The claim that *Ocean Shield* would find the plane turned out to be the worst April Fools' joke ever played. The mission was plagued with uncertainty from the moment the ship departed.

The formula used by Inmarsat to try to establish the plane's location was fraught with undefined variables and operational assumptions. The ocean floor in the remote destination of the Indian Ocean where they intended to search had never been mapped in detail. The area could be three thousand meters deep or maybe eight thousand meters deep—maybe more. No one had any idea what to expect. However, in spite of the doubts and criticisms, the desperation to do something, anything, to find the passengers and crew made the mission move forward.

Carrying a heavy cargo of hope and desperation, the vessel had been retrofitted during the five days leading up to the launch with towfish capable of hearing the 37.5 kHz pings that a ULB emits on contact with water. As in the search for Air France Flight 447, these devices were on loan from the US Navy. It was front and center in the investigators' minds that if the pings led them to the cockpit voice recorder and the flight data recorder, they'd know exactly what had led to the aircraft's demise.

One minor thing that helps investigators in their searches that the public is not usually aware of is that the CVR and FDR, the black boxes, are actually painted a bright orange, not black. This is done to make them more visible should investigators need to look for them in rubble or underwater. Some accident investigators believe the myth that they're referred to as black boxes because that's how they, like the rest of the plane, look after a fiery crash. Of course, most of us know that's not true. They're called black boxes because they were originally black, and the name stuck. The bright orange paint is heat resistant.

As the towfish moved about the ocean, they created images of the ocean floor and transmitted the images in real time to investigators. The towfish were attached to a cable that was deployed and retracted from an enormous winch mounted to the deck of the *Ocean Shield*. The winch was capable of reeling out as much as ten miles of cable, and as the *Ocean Shield* barreled through the rough sea, the towfish would be placed into the water and the cable would be slowly released from the winch. The more cable released, the deeper the towfish descended. As sophisticated as the system is, it is only capable of detecting a ULB's 37.5 kHz signals if it is within 1 to 5 kilometers, or about .6 to 3.1 miles, of the ULB.

The 37.5 kHz pings are about the pitch of a dog whistle and are far too high-pitched for the human ear to hear. If we could hear them underwater and stood close enough, the pings, which emanate almost every second, would perforate our eardrums. The scientific term for that intensity of sound is REALLY, REALLY LOUD.

In addition to the TPL system, the *Ocean Shield* had a Bluefin 21, an autonomous underwater vehicle (AUV) suitable for carrying out deep-sea missions, and a launch platform for it. The Bluefin-21 was designed to operate at depths up to 4,500 meters, or approximately 2.8 miles, and could be launched for detailed ocean floor searches with either photography or sonar equipment.

Ocean Shield deployed its first towfish on April 4, 2014, and its second towfish on April 5. The second towfish, while descending, detected a ping at a frequency of approximately 33 kHz. On April 8, pings were heard at around 27 kHz. This, as you can imagine, seemed like very big news. The investigators were in a race against time. They'd begun their mission on April 4 knowing that the ULB batteries of MH370 might only last thirty days, and the plane had already been missing for twenty-seven days. The burden of finding the plane before the ULB batteries died was a heavy weight to carry. The clock was ticking, and the world was watching.

On April 11, the Prime Minister of Australia, Tony Abbott, held a press conference in China, where 152 of MH370's 227 passengers were from, to discuss the pings. "We have very much narrowed down the search area, and we are very confident that the signals that we are detecting are from the black box[es] on MH370," he said. "We have a series of detections, some lasting for quite a long period of time."

As I listened to Abbott, I felt those all-to-familiar chills run down my spine. An authority, the Prime Minister of the country that was investing the most money into the search for MH370, had just announced that he was confident they had found the black boxes. With the black boxes, presumably, came the answer to what had happened to the aircraft. For an accident investigator, it's always a thrill when an aircraft has been found—and on a more personal level, I wanted the victims' families to get closure.

Oddly, just minutes after Abbott spoke, the Australian Search Coordinator, retired Air Chief Marshal Angus Houston, gave a contradictory statement. Houston said, "There has been no major breakthrough in the search for MH370." What did this mean? How could both statements make sense? If I, an experienced accident investigator and aviation crisis analyst, couldn't make sense of what was going on, I knew the families of the victims certainly couldn't— and again my heart went out to them.

9

EXPERIMENTING WITH KHZ

On May 12, 2014, just more than a month after Abbott and Houston spoke, Commander James Lybrand, Captain of the *Ocean Shield*, gave an interview to the *Wall Street Journal* that seemed to close the conversation. Lybrand said he felt the pings were too weak and of the wrong frequency to be man-made. "As far as frequency goes, between 33 kHz and 27 kHz is a pretty large jump," he said.

I heard this comment—and I respected Lybrand—but as an accident investigator, my job is to question everything. Even experts make mistakes, and a 33 kHz transmission, or even a 27 kHz one, didn't seem that far off from a 37.5 kHz one. Wasn't it possible that the ULB was simply not working as well as it should have? Wasn't it possible that damage was what led to a 33 kHz ping? And then a 27 kHz was produced by the ULB because its battery was fading? If these were possible, then maybe the transmissions were from the aircraft after all. I needed to confirm things myself.

As a little science experiment, I obtained several functional ULBs and set about testing the possible kHz transmissions that they could produce. My mother will attest to the fact that I have always enjoyed destructive experiments—many, she'll tell you, were when I was a kid playing with my sister's Barbie dolls, and others led to my blood, sweat, and tears. You'd think the blood, sweat, and tears ones—and the pain that came with them—would have taught me a lesson, but no, I still like science experiments as much as I did as a kid, and I still have a biting need to see proof for myself when I'm not 100 percent convinced of what I'm hearing. I consider this trait to be one of the reasons I've had a successful career. Others may jump to conclusions about why a plane crashed or went missing. I refuse to do so.

For the experiment, I needed to first see if a ULB was likely to have been damaged if the plane crashed—if it wasn't, then there was no point in moving forward because it should have kept transmitting a 37.5 kHz frequency until its batteries ran out.

I designed the experiment using mounting blocks to hold each end of a ULB in place. One end was stationary in the vise on my workbench, and the other was mounted to a four-inch long metal pipe. To get going, I produced a ULB's normal 37.5 kHz signal using a paperclip to short the activation switch. The paperclip tricked the ULB into producing the 37.5 kHz signal as if it were in water.

Next, I slowly but forcefully twisted the case of the ULB by pulling on the pry bar. I admit that this wasn't a very precise scientific experiment, so I have no idea how much twisting force I applied. I'll just say it was a lot—enough to make sweat run from my forehead into my eye and enough that the bar broke loose and I bashed my hand into the vise. (Again, blood, sweat, tears, and pain—lots of pain.) I pulled multiple times and each time was disappointed to see that despite the twists, there was no change in the frequency of the ULB's transmission. All I was doing was hurting myself. The ULB was probably laughing at me.

Next, I tried spraying the ULB with a freezing agent and super-heating it with a propane torch. Maybe the kHz transmission change that the *Ocean Shield* investigators heard was due to MH370's ULB being submerged in freezing cold water or being subjected to super heat from a fire. Again, these attempts did nothing to change the 37.5 kHz transmissions; it remained stable. By now, I was shouting at the ULB as I applied ice to my hand to relieve the pain from picking it up before the heat from the torch had dissipated. As my frustration built, I continued to bleed, my hand burned, and my eye stung. But persevere I would! I decided to advance to the next phase—the sledgehammer phase.

The sledgehammer phase was an even more archaic phase than the twist phase and the shout phase, in which I yelled at the ULB and the ULB ignored me. The sledgehammer phase consisted of my placing the ULB in a rubber cradle of sorts and banging on the unit with a sledgehammer to try to produce a change in its transmissions. The lack of results prompted me to advance from a one-pound hammer to a five-pound hammer, and with the first blow of the massive five-pound sledge finally came a change: the frequency nearly doubled, hitting 73 kHz! The feeling of accomplishment was short-lived, however, as I realized that the frequency did not go down as would have fit the theory that a damaged or aging ULB would produce lower-than-normal transmissions. There was no way that the first ULB transmission received by the *Ocean Shield* investigators would have been 33 kHz and then spiked.

In a fit of male obstinacy, I repeated this painful exercise two more times until I sat morose with three sad-looking, mutilated ULBs. Each time, I could produce unusually high transmissions, such as 73 kHz, but I could not produce lower ones. I was ready to give up, but to do so, I needed an explanation for the change in kHz signals received.

The answer came to me through Twitter.

Twitter, I've found, is a surprisingly civil social medium. Because it allows users to be anonymous, one might suspect it would attract the scourge of society, but it's quite the contrary. It has allowed millions of people who really care about the disappearance of MH370 and the families of the passengers and crew to express their thoughts and sentiments. A remarkable amount of people have given heartfelt prayers, and many have—in just 140 character or fewer statements—offered their expertise to help solve the mystery. Only a scant few chose to make disparaging comments and throw accusations at those trying to help. (A feature I like about Twitter is that it allows you to push those who choose the low road out of eyesight by clicking "BLOCK.")

Except myself, of course, the best accident investigators are usually a pain in the butt to deal with, but through Twitter I was introduced to an excellent accident investigator named John Fiorentino. John's instinct had also been that the pings could have come from MH370's ULB, but he made the case that they could also have come from other devices in that area of the ocean. This seemed reasonable, but to be convinced, I needed to know what other devices operate in the same frequency range as a ULB and also ping at one-second intervals. I explained this to John, and he assured me that he would provide the information I needed.

While I waited to hear from him, I reached out to the Woods Hole Oceanographic Institute (WHOI), a nonprofit institute that researches marine science and engineering and is the leading expert on oceanic tracking devices. During an investigative call with one of the WHOI scientists leading the institute's whale tracking efforts, I was told that they were aware of other devices that operate in the same frequency range as a ULB. However, the scientist said, the devices that operate in that frequency range and ping approximately every second are obsolete and have not been used in more than ten years.

At this point, I wasn't sure what to think. If there were no other devices in use that produce a frequency in the same range as a ULB—so something around 33 kHz or 27 kHz but not much lower or higher—didn't the pings detected have to have been from MH370's ULB? As promised, however, John came through and produced evidence to the contrary. He sent me information from the NASA Langley Research Center's National Aeronautics and Space Administration that showed me there are fish pingers, devices used to mark fishing nets, that operate within the same frequency range of a ULB and have a variable pulse setting of .5 to 2 seconds. I also came to learn that there are whales, dolphins, and other marine life that produce false pings, so Commander Lybrand may have been right on the money.

These discoveries shook the bedrock of my previous assumption about where the aircraft ended its flight, which was where the *Ocean Shield* had been searching. I now rated the Inmarsat data as less than 100 percent credible, particularly since it had never been used to locate an aircraft before. When it had seemed like the Inmarsat data was validated by the ULB pings that were heard in the ocean, the preponderance of "evidence" seemed to indicate that the aircraft would likely be found within days. When it turned out the Inmarsat data had not actually been validated by the pings that were heard, I felt I could no longer assume it to be accurate. What had seemed like proof that backed it up was not.

Because there are other devices in the ocean that produce sounds similar to those of a ULB, I highly recommend that ULBs be reconfigured to emit a more unique identifying pulse pattern—something such as Morse code that transmits the serial number of a unit. Without such a device, in figuring out what happened to the missing aircraft I was back at square one.

10

GEORESONANCE

On April 29, 2014, GeoResonance, an Australian company whose website describes it as a company that "offers a proven method of geophysical survey," issued a press release stating that on March 10, 2014, just two days after the disappearance of MH370, it found what it believed to be "wreckage of a commercial airliner" in the Bay of Bengal, a triangular body of water bordered by India, Sri Lanka, Bangladesh, Burma, and the Andaman and Nicobar Islands. The company said it found the wreckage by using a proprietary technology that scans vast areas for specific metals or minerals. The release included images that the company said were from 1,000 to 1,100 meters below sea level in the Bay of Bengal. The release read:

GeoResonance Pty Ltd ("GeoResonance"), a South Australian company and its team of scientists have invested considerable resources into the search for Malaysian Airlines flight MH370. The only motivation is to help the families of the missing passengers and crew, knowing the company has the technology capable of the task.

GeoResonance has discovered what they believe to be the wreckage of a commercial aircraft. The wreckage is located approximately 190 km south of Bangladesh in the Bay of Bengal. The wreckage is sitting on the seabed approximately 1000 m. to 1100 m. from the surface. The company is not declaring this is MH370, however, it should be investigated.

The search was completed using proven technology. In the past, it had been successfully applied to locate submersed structures, ships, munitions, and aircraft. In some instances, objects that were buried under layers of silt could not be identified by other means. At present the technology is being used with great success in the mining exploration industry.

In order to identify and locate subsurface substances, GeoResonance Remote Sensing analyses super-weak electromagnetic fields captured by airborne multispectral images. During the search for MH370, GeoResonance searched for chemical elements that make up a Boeing 777: aluminum, titanium, copper, steel alloys, jet fuel residue, and several other substances. The aim was to find a location where all those elements were present.

GeoResonance commenced the search before the official Search and Rescue operation moved to the

Southern Indian Ocean. The multi-discipline team of 23 researchers, including 5 professors and 12 PhDs got involved in the project. The search used the imagery taken on March 10, 2014, and was conducted consecutively in 4 zones north and northwest of Malaysia, until all targeted elements produced an anomaly in one place in the Bay of Bengal.

GeoResonance completed analysis of multispectral imagery of the location taken on March 5, 2014. It established that the anomaly had appeared between the 5th and 10th of March, 2014.

The approximate location was passed onto Malaysia Airlines and the Malaysian and Chinese Embassies in Canberra, Australia, on March 31, 2014. It was well before the black box batteries had expired. These details were also passed onto the Australian authorities (JACC) in Perth on April 4, 2014. A more detailed study was completed in early April. The final 23 page report including the precise location of the wreckage was passed onto Malaysia Airlines, Malaysian High Commission in Canberra, Chinese Embassy in Canberra, and the Australian authorities (JACC) on April 15, 2014.

The Company and its Directors are surprised by the lack of response from the various authorities. This may be due to a lack of understanding of the company's technological capabilities, or the JACC is extremely busy, or the belief that the current search in the Southern Indian Ocean is the only plausible location of the wreckage.

The people involved in the Channel 7 *Adelaide News* interview were Mr. Pavel Kursa, GeoResonance CEO, Mr. David Pope, GeoResonance Director, and

CEO of Tellus Resources, Mr. Carl Dorsch. Mr. Dorsch was involved as a client reference for the technology. The company directors are not seeking publicity, they only want to bring the results to the attention of the authorities. The directors feel a moral obligation to help bring closure for the families of the 239 passengers and crew of flight MH370 by releasing the findings, so the authorities can investigate.

The press release refers to technology the company uses that analyses super-weak electromagnetic fields captured by airborne multispectral images. Spectral remote sensing (SRS) has been in use for many years. I investigated the use of SRS for locating concentrations of extrusive igneous (volcanic) rock deposits for gravel mines in Route County, Colorado, in 2010, and my familiarity and respect for the technology prompted me to take the claims seriously. The Australian government's search team, the Joint Agency Coordination Centre (JACC), declined to investigate, however, saying, "The Australian-led search is relying on information from satellite and other data to determine the missing aircraft's location. The location specified by the GeoResonance report is not within the search arc derived from this data. The joint international team is satisfied that the final resting place of the missing aircraft is in the southerly portion of the search arc."

11

STOLEN PASSPORTS

When it was announced that two Iranian passengers of MH370, nineteen-year-old Pouria Nourmohammadi Mehrdad and twenty-nine-year-old Seyed Mohammed Reza Delavar, booked their tickets using stolen passports, the investigation leaned dramatically toward suspicion of nefarious activity. People around the world immediately thought of the 9/11 hijackers, and the two men were suspected of being terrorists. Among the rumors flying were that the men killed everyone on board the plane, escaped, and let it crash or that they took the plane for their own personal use or for the use of a terrorist group they were part of. Maybe, some of the rumors said, they took the plane to be retrofitted with bombs and used in a 9/11-style attack on the United States. I wasn't above these rumors. It seemed possible to me that one of the men had taken control of the aircraft while the other had somehow killed or subdued the passengers and crew. As an American and as an accident investigator, the 9/11 attacks are always in the back of my mind.

Tempers flared with authorities in Tehran when the connection to Iran was suspected. A prominent Iranian lawmaker, Seyed Hossein Naghavi Hosseini, the spokesman for Iran's Parliament National Security and Foreign Policy Committee, lambasted the theories, telling an Iranian news agency, the Tasnim News Agency, that the rumors were equivalent to "psychological warfare." According to Hosseini, "Americans recruit some people for such kinds of operations so they can throw the blame on other countries, especially Muslim countries."

A spokesperson for the Iranian Ministry of Foreign Affairs, Marzieh Afkham, seemed more grounded in her response to the news. "We have received information on the possible presence of two Iranians [aboard the plane and] . . . are pursuing the issue. We have informed our embassy in Malaysia that we are ready to receive further information about the issue from Malaysian officials. We have announced that we were ready for cooperation."

In a news conference on March 11, 2014, Khalid Abu Bakar, the Inspector General of the Malaysian police, said the younger of the two Iranian suspects,

Pouria Nourmohammadi, had been cleared of suspicion. He said Nourmohammadi was using a stolen Austrian passport to travel to Germany to meet his mother, who was waiting for him in Frankfurt, and that the Malaysian police were in contact with his mother. Later, in May 2014, Nourmohammadi's mother, Niloufar Vaezi Tehrani, through an interpreter, told News Corps Australia that Nourmohammadi was just an "ordinary kid. . . . Judging people is very easy," she said, "but people don't know the truth. Pouria was an ordinary kid. He was a student at university . . . and he just had a normal life . . . It's not fair to push and point everything regarding this incident at Pouria. They need to look at what was the actual and real reason for this incident."

Seyed Mohammed Reza Delavar was using a stolen Italian passport, and he, too, was cleared of suspicion by the Malaysian police. It's thought that both men were trying to immigrate to Europe—Nourmohammadi to Germany and Delavar to Denmark, which was his final ticketed stop.

In a weird turn of events, it came out that photos circulating of the two men had been altered. Nourmohammadi and Delavar, who seemed to be traveling together, have identical legs and shadows in the photos. This was noted by another smart Twitter user. A spokesman for the Malaysian police confirmed the photos were messed up but said it "was not done with malice or to mislead." I'm not sure what to make of this. I mention it to emphasize that authorities do make mistakes—the reason why I felt it important to investigate for myself what might have happened to MH370 instead of taking for granted what the authorities are saying happened to the plane.

After the Iranian men were cleared, the part of the investigation considering whether nefarious activity may have led to the plane's downfall shifted back toward the pilots.

12

THE PILOTS

The evidence of MH370 making its initial and subsequent turns off course is conclusive that either a person was flying the aircraft and led the plane off course, or that a person entered an off-course flight plan into the flight management computer. The story of MH370 may seem like it's out of some sci-fi show, but this isn't *Star Trek*. A plane doesn't decide by its own volition that it wants to journey off course.

Given this, it's not at all surprising that the pilots of MH370 were suspected of being responsible for something terrible happening to it. Any time an aircraft goes missing or crashes, its pilots are going to be suspects since they have the most control over the plane and where it goes, and this will bring them under close scrutiny. Questions about their backgrounds and mental stabilities will be asked. Among them, is there any reason they'd want to destroy an aircraft? And is there any reason they'd want to harm themselves or harm others?

It's commonly thought that there have been three commercial airline pilot suicides since 1997. The earliest concerns the pilot of SilkAir Flight 185, whose Boeing 737 crashed on December 19, 1997, killing all 104 people on board. The aircraft was supposed to go from Jakarta, Indonesia, to Singapore but crashed in Indonesia after a rapid descent from its cruising altitude. The plane plunged downward so quickly—faster than the speed of sound—that it broke apart in the process, its debris spreading over several kilometers. It was later determined that the plane's first officer had left the flight deck just before the aircraft began its disastrous descent. It was theorized that the pilot waited until his copilot left the cockpit so he could easily commandeer the plane.

EgyptAir Flight 990 is the second of the three that are thought to have crashed as a result of pilot suicide. Flight 990 crashed into the Atlantic Ocean on Halloween, October 31, 1999, and claimed the lives of all 217 people on board. The aircraft, a Boeing 767, made a rapid descent that led to its crash just thirty minutes after its departure from New York's John F. Kennedy International Airport. The National Transportation Safety Board concluded that the tragedy was the "result of the relief first officer's flight control inputs," which were input just after the pilot left the flight deck and that "the reason for the relief first officer's actions was not determined." The suggestion that a deliberate act led to the crash was heavily disputed by Egyptian authorities.

The third crash thought to be the result of pilot suicide is that of LAM Mozambique Airlines Flight 470. It took place on November 29, 2013, and claimed the lives of the aircraft's thirty-three passengers and crew members. The aircraft, an Embraer 190, was supposed to fly from Maputo, Mozambique, to Luanda, Angola, and the flight should have taken just four hours, but of course, it never made it. The plane made a rapid descent and crashed in Bwabwata National Park in northern Namibia in southern Africa. Once again, the control inputs that led to the plane's demise were entered shortly after the first officer left the flight deck.

I am not a psychologist and will not profess to know the mental state of the pilots of these plane crashes nor of Captain Zaharie Ahmad Shah or First Officer Fariq Abdul Hamid of MH370. That being said, I was led to my opinion that neither Shah nor Hamid desired to hurt himself or the passengers of MH370 by descriptions of them given by their families and friends. The calm, collected nature of their voices in the recordings on board the aircraft as they spoke to ATC, which I mentioned earlier, reinforces this belief.

The sister of Captain Zaharie, Sakinab Ahmad Shah, spoke about her brother, a father of three, to Channel NewsAsia, an Asian TV station, for a special feature they aired on July 13, 2014, titled *The Mystery of MH370*. In the feature, Zaharie's sister describes him by saying, "He loved aviation; he spent a lot of his funds buying model airplanes. If he could, I think he would attach wings to himself and fly—he loved flying that much. . . . If it was done, if he was the one who planned it, he has to be some kind of Einstein, which he was not." (I think this was said with love.)

Pictures that have circulated on social media of Captain Zaharie include ones that show him cooking a traditional rice vermicelli soup dish called bihun, a tofu sambal, and a fish curry, as well as him on a fishing trip. His Facebook page shows meals he's made and has photographs of him eating with family members and also flying a miniature airplane. In all the pictures, Zaharie seems like a happy, normal person. And since MH370 disappeared, his nieces and nephews have posted on Twitter praying for his safe return home. Although it seems that Zaharie did have firm political beliefs—it was said that he was a "fanatical supporter" of Anwar Ibrahim, Malaysia's opposition leader—there have not been accounts of him being violent or stating intent to use violence to make a political stand. And Ibrahim is not an encourager of violent actions.

By all accounts, twenty-seven-year-old Fariq Abdul Hamid was also a happy, normal person—also far from someone who'd want to kill himself and 238 other people. I spoke to two people who knew Fariq and was told that he had a promising future with Malaysia Airlines and was planning to marry Nadira Ramli, a twenty-six-year-old who flies for AirAsia. At the time of Fariq's disappearance, the two had known each other for nine years. They reportedly met at the Langkawi Aerospace Flying Academy in Malaysia, where they bonded over a love of aviation and often studied together.

At present, it seems that investigators have ruled out First Officer Hamid as a suspect in the plane's disappearance but that Zaharie remains under investigation. One reason for this could relate to the curious nature of a sophisticated flight simulator that was found in his home. The simulator allows its users to practice various routes, and questions have arisen as to why Zaharie needed to have such a device in his home when the Malaysia Airlines Flight Crew Training Centre Complex in nearby Subang, which also had flight simulators, was available to him at any time. One speculation is that Shah wanted to practice alternate flight routes than the one MH370 was supposed to take. Fuel was added to this fire when it was revealed that the simulator's hard drive was sent to FBI laboratories for expert analysis and that the

FBI determined that files had been removed from the hard drive. Perhaps, some thought, Zaharie practiced an alternative course and then deleted the files after, knowing investigators would raid his home after the plane went missing. This theory, however, is wild speculation. The FBI, after searching the simulator, reported that they found "nothing suspicious whatsoever" on the hard drive, and since the news about the hard drive broke, other pilots have come forward in defense of Zaharie having it.

Malaysia Airlines flight MH370 is scheduled to arrive in Beijing on March 8, 2014 at 6:30 a.m. Families of passengers wait to greet their loved ones at the airport. The first indication they have that something is wrong is the red lettering on the arrival board in the main terminal at the Beijing Capital International Airport.

There is no announcement about the flight over the public address system at the airport. There is no one in the terminal with information about the flight. There is just silence as the families anxiously await news about when the aircraft will land. (KIM KYUNG-HOON/Reuters/Corbis)

Malaysian Prime Minister Najib Razak, *front left,* and Australian Prime Minister Tony Abbott, *front right,* attend a briefing at the Royal Australian Air Force base, Pearce, in Perth, Australia, near where searchers are looking for the missing aircraft, on April 3, 2014. At the invitation of Abbott, Razak stays at the air force base for two days. (RICHARD WAINWRIGHT/epa/Corbis)

A Malaysian police official displays photographs of the two Iranian men, Pouria Nour Mohammad Mehrdad and Seyed Hamid Reza Delavar, who boarded Flight 370 with stolen passports.

The announcement that two men boarded the plane with stolen passports reveals yet another vulnerability in international travel, that many countries including Malaysia do not check passengers' passports against the International Police Organization's (INTERPOL's) Stolen and Lost Travel Documents database. The database contains information on more than forty million lost travel documents. (Adli Ghazali/Corbis)

MH370's pilot, Zaharie Ahmad Shah, at home with his personal flight simulator. There are theories that pilot suicide or political motivations led Shah to commandeer the aircraft and crash it into the Indian Ocean. These theories find no support following careful examination of the simulator's hard drives and the backgrounds of both the pilot and his co-pilot, Fariq Abdul Hamid, by the FBI, these theories. (Mirrorpix/Splash News/Corbis)

During a press conference at Kuala Lumpur International Airport on March 22, 2014, Malaysia's acting transport minister, Hishammuddin Hussein, is interrupted and handed a note. He immediately relays its information to the public. It reads that Chinese satellites have spotted objects floating in the southern part of the search area for the missing aircraft; that the objects could be debris from the plane; and that ships have been sent to investigate. The objects prove to be nothing more than trash floating at sea. (EDGAR SU/Reuters/Corbis)

From left: Air Chief Marshal Angus Houston, chief coordinator of the Joint Agency Coordination Centre, Hishammuddin Hussein, Malaysia's acting transport minister, Jean-Paul Troadec, president of the French Aviation Accident Investigation Bureau, and another man walk to attend a news conference on the missing aircraft on May 2, 2014 in Kuala Lumpur, Malaysia.

The Malaysian government reaches out to the Australian government for assistance in the search for MH370, and Houston provides calm, confident leadership. (SAMSUL SAID/Reuters/Corbis)

Erin Gormley, an aerospace engineer at the National Transportation Safety Board, dips an underwater locator beacon removed from a flight recorder into water to show how it emits a ping on contact. If searchers succeed in the Herculean task of retrieving the black boxes (actually orange in color as the photo shows) from the depths of the Indian Ocean, a new recovery job will begin to retrieve information from the flight data and voice recorders. (Charles Dharapak/AP/Corbis)

A US Navy towed pinger locator towfish on a dock at Her Majesty's Australian Ship (HMAS) Stirling, a naval base on the west coast of Australia, near Perth, on March 30, 2014. The drone, capable of exploring waters nearly 15,000 feet deep, is towed behind the Australian Defense vessel *Ocean Shield* in a search for the aircraft's black boxes. Almost immediately after it is put into the Indian Ocean, it begins picking up signals that seem to be from an underwater locator beacon. It later turns out that they are unrelated. (JASON REED/Reuters/Corbis)

Ocean Shield carrying the Bluefin-21 submersible underwater drone at HMAS Stirling on May 10, 2014, before it leaves on its second mission to scan part of the Indian Ocean floor where the longest sonar ping thought to be from the black box was heard nearly a month earlier. (STRINGER/Reuters/Corbis)

The Bluefin-21 sits on the wharf ready to be fitted to *Ocean Shield* on April 14, 2014. Based on analysis of the acoustic detections underwater sonar, the submersible underwater drone will use its side-scan sonar to search at depths between 3,800–5,000 meters at a ten kilometer radius area near where the most promising ping was earlier detected by the towed pinger locator and a 3 km radius area around three other promising locations.

The search lasts nearly six weeks and is completed on May 28, 2014. The total area searched is 860 km2. Nothing is found. On its completion, the Australian Transport Safety Bureau announces that the area is no longer considered a possible final resting place for MH370. (Rob Griffith/AP/Corbis)

On June 26, 2014, following an examination of MH370's flight path which determined that the turns made by the aircraft were commanded by human actions, Australian Deputy Prime Minister Warren Truss shows the new search area that will be covered. It is still not known if the commands were input through the flight controls or if they were preprogrammed into the system and the plane flew the course on autopilot. (ALAN PORRITT/epa/Corbis)

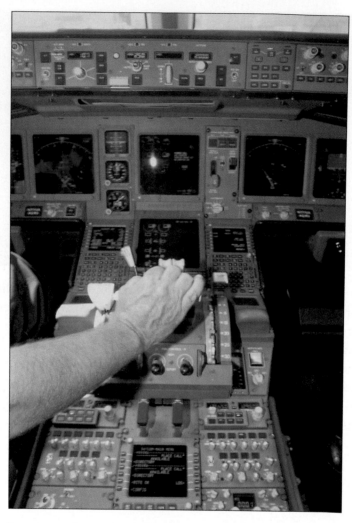

A pilot manipulates the flight controls of a Boeing 777, the same model aircraft as MH370. The aircraft's flight management system controls are on the top of the glare shield and by twisting the heading control to the left or right, they direct the aircraft. (Kim Kulish/CORBIS)

At their London headquarters on March 25, 2014, staff at satellite communications company Inmarsat work in front of a screen showing subscribers using their service throughout the world,

Inmarsat uses a wave phenomenon, called Doppler, to analyze the seven pings its satellite picked up from MH370. Its findings are what led Malaysian Prime Minister Najib Razak to conclude that the Boeing 777 crashed thousands of miles off course in the southern Indian Ocean. (ANDREW WINNING/Reuters/Corbis)

Malcolm Johnson, director of the International Telecommunication Union, and Nancy Graham, director of the International Civil Aviation Organization, attend a news conference after an expert meeting regarding real-time tracking for commercial aircraft at a hotel in Kuala Lumpur on May 27, 2014.

The call for in-flight tracking of aircraft and the streaming of black box data was a recommendation following the two-year search and recovery efforts for Air France Flight 447. (SAMSUL SAID/Reuters/Corbis)

A member of Inmarsat works in front of a screen showing subscribers using their service throughout the world, at their headquarters in London on March 25, 2014. (ANDREW WINNING/Reuters/Corbis)

Malaysia's acting transport minister, Hishammuddin Hussein (center), the deputy minister of its Department of Foreign Affairs, Hamzah Zainuddin (left), and the director general of its Department of Civil Aviation, Azharuddin Abdul Rahman (right), look at a note given to Hussein during a press conference on MH370. (Fazry Ismail/NurPhoto/Corbis)

13

RUMORS

One of the earliest unsubstantiated rumors about what might have happened to MH370 started with Malaysian officials. The officials reported there were five passengers who purchased tickets for the flight but did not board the aircraft. Khalid Abu Bakar, the Inspector General of the Malaysian police, later reported this was false.

The real hot spot for rumors is, of course, the Internet. As I said earlier, the Internet can be a great help in solving mysteries. There are so many experts who share their knowledge on it. But people also make all kinds of claims without accountability. And, sadly, there are people who buy into the ridiculous theories they read.

One rumor that started online was related to the fact that twenty passengers on MH370 were employees of Freescale Semiconductor, Inc., a company that designs and produces embedded hardware and software. Twelve people on board the aircraft worked in the company's Kuala Lumpur branch, and eight people worked in its Tianjin, China, branch. This was confirmed by a spokesperson for the company.

What is not true, however, is the wild, highly imaginative theory circulated online about the MH370 aircraft being confiscated or purposefully destroyed because of this. One anonymous contributor to the site *Before It's News* was particularly aggressive in spewing this theory and linked the disappearance of the plane to what was described as radar-blocking capabilities of aeronautical hardware technology produced by Freescale. "It is conceivable that the Malaysia Airlines Flight MH370 plane is 'cloaked', hiding with high-tech electronic warfare weaponry that exists and is used," the writer typed. Well, of course . . . NOT! Where does this stuff come from?

And wait, there's more!

The writer also promulgated another conspiracy theory, which, too, starts with facts then falls into fantasy. The theory accurately reports that some Freescale employees are listed on a US patent, 8671381 B1, which was assigned to the company on March 11, 2014, three days after MH370 vanished. It then goes on to claim that four of the five names on the patent, not including Freescale's own name, were on board MH370. The four people it names are Peidong Wang, Zhijun Chen, Cheng (no first name provided), and Li Ying Zhijong. One version of the theory even makes an almost-convincing claim that Freescale Semiconductor is owned by The Blackstone Group, which it says is owned by the British billionaire Jacob Rothschild. With four of the five patentees dead, the patent would fall into the cunning hands of Rothschild. This sounds like the makings of a great movie, doesn't it? Too bad! Rothschild is a member of Blackstone's International Advisory Board, not its owner, and more importantly, a fact check shows that none of the four patent members mentioned were actually on the aircraft. I've included a passenger list in the next chapter if you want to see this for yourself. The list, I should add, is publicly accessible. Anyone could pull it up online and see that none of the four people mentioned were on the missing aircraft. This theory, by the way, does not accurately reflect how a patent works. Unless other provisions are made, the share of the patent that belongs to a deceased owner is passed to his or her heirs, not given to someone else on the patent.

There are a great many more wild conspiracy theories about what happened to MH370. I won't get into them, and I admit I haven't investigated every one of them. There simply is not enough time to do so—and there is nothing to gain by doing so.

14

THE PASSENGERS

The decision to include a passenger list and their seat locations in this book was not an easy one. The idea of doing so weighed heavy on my heart. I feature the list here in order to provide a more intimate connection to the passengers and crew. Each time I read the list, it brings me closer to comprehending the enormity of the tragedy. Somber reflection upon the day the flight went off the radar and the deep loss I feel motivates me to continually strive to influence improvements in aviation safety.

Passengers from two to seventy-nine years of age were on MH370. Their nationalities were from fifteen countries. On Saturday, March 8, 2014, after they boarded the plane, they put their belongings in the overhead bins, took their seats, buckled their seatbelts, listened to the flight attendant briefing them in several languages, and prepared themselves for a six-hour flight to Beijing. They ate snacks and sipped beverages as their journey began. Most likely, they read, chatted with one another, and listened to music. As far as anyone can tell, the flight got off to a normal start.

Official List of MH370 Passengers and Seating Plan:

Name	Nationality	Age	Gender	Seat No.
GAN/FUXIANG	CHN	49	M	2A
CHNG/MEI LING	MYS	33	F	2D
WEEKS/PAUL	NZL	39	M	2K
YUE/GUI JUMS	CHN	51	F	3A
BRODSKII/NIKOLAI	RU	43	M	3K
XING/QIAO	CHN	27	F	4K
CHE/JUNZHANG	CHN	68	F	4J
MOHAMAD SOFUAN/ IBRAHIM	MYS	33	M	4J
MUKHERJEE/ MUKTESH	CAN	42	M	4C
BAI/XIAOMO	CAN	37	F	4A
TIAN/QINGJUN	CHN	51	M	11A
WOOD/PHILIP	USA	51	M	11C
LIN/ANNAN	CHN	27	M	11D
XIE/LIPING	CHN	51	F	11E
ZHANG/XUEWEN	CHN	61	M	11H
HUE/PUI HENG	MYS	66	M	11J
LI/LE	CHN	36	M	11K

Name	Nationality	Age	Gender	Seat No.
BIBY NAZLI/MOHD HASSIM	MYS	62	F	12A
DINA/MOHAMED YUNUS RAMLI	MYS	30	F	12C
MARIA/MOHAMED YUNUS RAMLI	MYS	52	F	12D
HASHIM/NOORIDA	MYS	57	F	12E
WANG/SHOUXIAN	CHN	69	M	12G
KOH/TIONG MENG	MYS	40	M	12H
MUHAMMAD RAZAHAN/ZAMANI	MYS	24	M	12J
NORLIAKMAR/ HAMIDMDM	MYS	33	M	12K
ZHANG/SHAOHUA	CHN	32	F	14A
ZHANG/HUALIAN	CHN	42	F	14C
LI/YUCHEN	CHN	27	M	14D
LI/YUAN	AUS	33	M	14E
LI/JIE	CHN	27	F	14F
KANG/XU	CHN	34	M	14G
GU/NAIJUN	AUS	31	F	14H
SHARMA/ CHANDRIKA	IND	51	F	14J
CHAN/HUAN PEEN	MYS	46	M	14K
LEE/SEWCHUMDM	MYS	55	F	15A
NG/MAYLIMS	MYS	37	F	15C
YANG/LI	CHN	35	F	15D
LU/JIANHUA	CHN	57	M	15E
MA/WENZHI	CHN	57	F	15F
SIM/KENGWEI	MYS	53	M	15G

Name	Nationality	Age	Gender	Seat No.
ZHANG/CHI	CHN	58	F	15H
WAN/HOCK KHOON	MYS	42	M	15J
TEE/LIN KEONG	MYS	50	M	15K
LI/ZHIJIN	CHN	30	M	16A
YAP/CHEE MENG	MYS	39	M	16C
SUGIANTO/LO	IDN	47	M	16D
VINNY/ CHYNTHYATIO	IDN	47	F	16E
ZHANG/QI	CHN	31	F	16G
BIAN/LIANGJING	CHN	27	M	16H
MAT RAHIM/ NORFADZILLAH	MYS	39	F	16J
TONG/SOON LEE	MYS	31	M	16K
WENG/MEI	CHN	39	F	17A
ZHANG/NA	CHN	34	F	17D
HU/XIAONING	CHN	34	M	17E
HU/SIWAN	CHN	3	F	17F
SIREGAR/FIRMAN CHANDRA	IDN	25	M	17G
SHIRSATH/KRANTI	IND	44	F	17H
MUSTAFA/SUHAILI	MYS	31	F	17J
LEE/KAH KIN	MYS	32	M	17K
GAN/TAO	CHN	44	M	18A
LU/XIANCHU	CHN	33	M	18C
ZHAO/QIWEI	CHN	37	M	18D
ZHANG/XIAOLEI	CHN	32	F	18E
ZHAO/YINGXIN	CHN	3	F	18F
LIU/QIANG	CHN	40	M	18H

Name	Nationality	Age	Gender	Seat No.
RAMLAN/SAFUAN	MYS	32	M	18J
YUSOP/MUZI	MYS	50	M	18K
JU/KUN	CHN	32	M	19A
WANG/XIMIN	NZL	50	M	19C
SUADAYA/FERRY INDRA	IDN	42	M	19D
SUADAYA/HERRY INDRA	IDN	35	M	19E
TANURISAM/INDRA SURIA	IDN	57	M	19F
WANG/ WILLYSURIJANTO	IDN	53	M	19G
YANG/AILING	CHN	60	F	19H
GUAN/HUAJIN	MYS	34	F	19J
JIA/PING	CHN	32	F	19K
TAN/CHONG LING	MYS	48	M	1C
BURROWS/RODNEY	AUS	59	M	20A
BURROWS/MARY	AUS	54	F	20C
CHEN/JIAN	CHN	58	M	20D
FENG/JIXIN	CHN	70	M	20F
WANG/YONGGANG	CHN	27	M	20G
ZHANG/JIANWU	CHN	31	M	20H
MENG/GAOSHENG	CHN	64	M	20J
DING/LIJUN	CHN	43	M	20K
LAWTON/ CATHERINE	AUS	54	F	21A
LAWTON/ROBERT	AUS	58	M	21C
XING/FENGTAO	CHN	36	M	21D
CHEW/KAR MOOI	MYS	31	F	21E

Name	Nationality	Age	Gender	Seat No.
WONG/SAI SANG	MYS	53	M	21F
LIM/POWCHUA	MYS	43	M	21G
TAN/AH MENG	MYS	46	M	21H
TAN/WEI CHEW	MYS	19	M	21J
CHUANG/HSIU LING	TWN	45	F	21K
GUAN/WENJIE	CHN	35	M	22A
LIU/JINPENG	CHN	33	M	22C
WANG/CHUNHUA	CHN	34	M	22D
ZHANG/SIMING	CHN	71	F	22H
DOU/YUNSHAN	CHN	61	M	22J
ZHANG/LIJUAN	CHN	61	F	22K
XIN/XIXI	CHN	32	F	23A
HUANG/YI	CHN	30	F	23C
LIU/ZHONGFU	CHN	72	M	23D
MAIMAITIJIANG/ ABULA	CHN	35	M	23E
MAO/TUGUI	CHN	72	M	23F
ZHAO/GANG	CHN	46	M	23H
WEN/HAO DONG	CHN	32	M	23J
YAN/XIAO DAN	CHN	27	F	23K
OUYANG/XIN	CHN	38	F	24A
YIN/BOYAN	CHN	33	M	24C
ZHANG/YANHUI	CHN	44	F	24D
YANG/JIABAO	CHN	26	F	24E
WANG/DAN	CHN	54	F	24F
LI/HONGJING	CHN	20	F	24G
LI/GUOHUI	CHN	56	M	24H
HUANG/TIANHUI	CHN	43	M	24J

Name	Nationality	Age	Gender	Seat No.
JIANG/CUIYUN	CHN	62	F	24K
LIU/SHUNCHAO	CHN	46	M	25A
DI/JIABIN	CHN	36	M	25C
WATTRELOS/AMBRE	FRA	14	F	25D
WATTRELOS/ HADRIEN	FRA	17	M	25E
WATTRELOS/ LAURENCE	FRA	52	F	25F
ZHAO/YAN	FRA	18	F	25G
TANG/XUDONG	CHN	31	M	25H
YAN/LING	CHN	29	M	25J
CHEN/CHANGJUN	CHN	35	M	25K
ZHAO/PENG	CHN	25	M	26A
TIAN/JUNWEI	CHN	29	M	26C
LUI/CHING	CHN	45	F	26D
WANG/SHU MIN	CHN	61	F	26E
WANG/XIANJUN	CHN	61	M	26F
LI/ZHIXIN	CHN	35	M	26H
TAN/SIOH PENG	MYS	42	F	26J
CHEN/WEI HOING	MYS	43	M	26K
DEINEKA/SERGII	UKR	45	M	27D
CHUSTRAK/OLEG	UKR	45	M	27E
ZANG/LINGDI	CHN	58	F	27G
LIANG/LUYANG	CHN	60	M	27H
MOHD KHAIRUL AMRI/SELAMAT	MYS	29	M	29A
PUSPANATHAN/ SUBRAMANIAM	MYS	34	M	29C

Name	Nationality	Age	Gender	Seat No.
JIANG/XUEREN	CHN	62	M	29J
LI/YANLIN	CHN	29	M	29K
JIANG/YING	CHN	27	F	30A
KOZEL/CHRISTIAN[1]	AUT	30	M	30C
XU/CHUANE	CHN	57	M	30D
ZHANG/YAN	CHN	36	F	30E
MENG/BING	CHN	40	M	30F
MENG/FANQUAN	CHN	70	M	30G
MENG/NICOLE	USA	4	F	30H
ZHANG/MENG	CHN	29	F	30J
YAN/PENG	CHN	29	M	30K
ZHOU/JINLING	CHN	61	M	31A
ZHOU/FENG	CHN	56	F	31C
ZHAO/ZHAO FANG	CHN	73	F	31D
XIONG/DEMING	CHN	63	F	31E
WANG/LINSHI	CHN	59	M	31F
LOU/BAOTANG	CHN	79	M	31G
LIU/RUSHENG	CHN	76	M	31H
DONG/GUOWEI	CHN	48	M	31J
BAO/YUANHUA	CHN	63	F	31K
CHEN/YUN	CHN	57	F	32A
DING/YING	CHN	62	F	32C
HOU/AIQIN	CHN	45	F	32D
SONG/CHUNLING	CHN	60	F	32E
TANG/XUEZHU	CHN	57	F	32F
YANG/QINGYUAN	CHN	57	M	32G

[1] Pouria Nourmohammadi Mehrdad (IRA, 19, M) was traveling with Christian Kozel's passport.

Name	Nationality	Age	Gender	Seat No.
YANG/XIAOMING	CHN	59	F	32H
SIMANJUNTAK/ SURTI DAHLIA	NLD	50	F	32J
FENG/DONG	CHN	21	M	32K
CAO/RUI	CHN	32	F	33A
MA/JUN	CHN	33	M	33C
SONG/FEIFEI	CHN	32	M	33D
ZHANG/HUA	CHN	43	M	33E
TEOH/KIMLUN	MYS	36	M	33G
YAO/LIFEI	CHN	31	M	33H
YA/NA	CHN	26	F	33J
HAN/JING	CHN	53	F	33K
FU/BAOFENG	CHN	28	M	34A
MARALDI/LUIGI[2]	ITA	37	M	34C
YUE/WENCHAO	CHN	26	M	34D
ZHANG/YAN	CHN	45	F	34G
WANG/YONGQIANG	CHN	30	M	34H
SU/QIANGGUO	CHN	71	M	34J
SONG/KUN	CHN	25	M	34K
KOLEKAR/VINOD	IND	59	M	35A
KOLEKAR/CHETANA	IND	55	F	35C
KOLEKAR/SWANAND	IND	23	M	35D
WANG/HAITAO	CHN	26	M	35G
LUO/WEI	CHN	29	M	35H
LI/YAN	CHN	31	F	35J
GAO/GE	CHN	27	F	35K

[2] Seyed Mohammed Rezar Delawar (IRA, 29, M) was traveling with Luigi Maraldi's passport.

Name	Nationality	Age	Gender	Seat No.
HOU/BO	CHN	35	M	36A
WEN/YONGSHENG	CHN	34	M	36B
WANG/CHUNYONG	CHN	43	M	36C
LI/ZHI	CHN	41	M	36G
TAN/TEIK HIN	MYS	32	M	36H
DU/WENZHONG	CHN	50	M	36J
LIN/MINGFENG	CHN	34	M	36K
SHI/XIAN WEN	CHN	26	M	37A
WANG/LIJUN	CHN	49	M	37C
WANG/RUI	CHN	35	M	37D
JIAO/WEIWEI	CHN	32	F	37H
ZHOU/SHIJIE	CHN	64	M	37J
YIN/YUEWANG	CHN	21	M	37K
LI/WENBO	CHN	29	F	38A
ZHENG/RUIXIAN	CHN	42	F	38C
JIAO/WENXUE	CHN	58	M	38D
JINGHANG/JEE	MYS	41	M	38G
DAI/SHULING	CHN	58	F	38H
BIAN/MAOQIN	CHN	67	F	38J
WANG/YONGHUI	CHN	33	M	38K
LI/MINGZHONG	CHN	69	M	39A
LIU/FENGYING	CHN	65	F	39D
YUAN/JIN	CHN	63	F	39G
ZHANG/LIQIN	CHN	43	F	39J
WANG/HOUBIN	CHN	28	M	39K
ZHU/JUNYAN	CHN	41	F	40A
ZHANG/ZHONGHAI	CHN	43	M	40C
ZHANG/JINQUAN	CHN	72	M	40D
YAO/JIANFENG	CHN	70	M	40E

Name	Nationality	Age	Gender	Seat No.
YANG/MEIHUA	CHN	65	F	40F
DAISY/ANNE	MYS	56	F	40G
AN/WENLAN	CHN	65	F	40K
LIANG/XUYANG	CHN	30	M	41D
DING/YING	CHN	28	F	41G
WANG/MOHENG	CHN	2	M	34H
ZHANG/YAN	USA	2	M	34G

Passengers by Nationality:

Australia: 6

Canada: 2

China: 152

France: 4

Hong Kong: 1

India: 5

Indonesia: 7

Iran: 2

Malaysia: 50

Netherlands: 1

New Zealand: 2

Russia: 1

Taiwan: 1

Ukraine: 2

United States: 3

Total: 239

15

WHO PAYS?

On November 4, 2014, Hugh Dunleavy, Malaysia Airlines' Director of Commercial, told the *New Zealand Herald* that the Australian and Malaysian governments would likely set a formal date for declaring Flight MH370 lost by the end of 2014. Dunleavy said that once the plane was formally declared lost, Malaysia Airlines would begin the process of financially compensating the victims' families for the loss of their loved ones. Not surprisingly, Dunleavy's statement was met with shock and grief by families of the passengers and crew who continued to hope the plane would be found and didn't want the search to end prematurely. A family support group for the victims' families, Voice370, said that Dunleavy's comments added further "blows and heartbreaks" to those who'd lost their loved ones and that any suggestion that the plane would be prematurely declared lost brought "intense agony and confusion to family members" and made them "lose faith in the search efforts."

Australia's JACC quickly responded to this, saying they found Dunleavy's comments "greatly disturbing." In a statement, they said, "Australia continues to lead the search for MH370 on behalf of Malaysia and remains committed to providing all necessary assistance in the search for the aircraft." Following this, Malaysia Airlines did immediate damage control, declaring the comments "an expression of a personal opinion only" and saying the search would not be abandoned without the recommendation of the JACC.

Dunleavy's statement, while viewed as insensitive, was right on the money, and what he said will need to be discussed sooner or later. Once the plane is formally declared lost, the process of compensating the victims' families begins. This compensation process reflects those outlined in Malaysia Airlines' insurance policies, and its policies reflect industry-standard ones.

Commercial aviation insurance works like this: To be considered properly insured, airlines must either prove they can self-insure or purchase several different insurance products. The primary product or form is hull loss, which pays for an aircraft if it is damaged beyond repair or not recoverable. The second form is liability insurance, which helps airlines make payments to passengers or anyone on the ground who is harmed or killed by an aircraft. The third form is war risk, which pays when a plane is hijacked, otherwise destroyed by terrorists, or shot down by a foreign air force.

Collecting payment from liability insurance before 1999 was a tangled web. Each country and each insurance policy had different rules about settlements. The Montreal Convention of 1999 (MC99), a multilateral treaty adopted by ICAO and its member states, changed all that for airlines and their passengers. It was a much-needed modernization and unification of what was formerly a hodgepodge of assorted aircraft liability treaties thrown together for the seventy years since the Warsaw Convention. With MC99, surviving family members are no longer forced to fight through the spaghetti of legal trappings to be compensated.

Since nearly all airlines worldwide, including Malaysia Airlines, have signed MC99, there is greater certainty about the rules affecting their liability, and passengers have a more streamlined handling and payment process for their claims. There is a clear rule that surviving family members receive a minimum of 150,000 dollars in the event that one of their loved ones is killed. And this does not restrict further legal action by the families.

The form of Malaysia Airlines' insurance policy that is war risk insurance is the source of much consternation about the fate of MH370 among the airlines' management, insurance company, and investigators. If it is confirmed that terrorist-related actions caused the loss of MH370, then the war risk insurers, led by a unit of Lloyd's of London, a London insurance market, would share in paying for a large portion of liabilities associated with the lost plane. However, without the aircraft's black boxes, which would supply the needed proof of this, they'll likely be off the financial hook.

On March 14, 2014, Credit Suisse sent a statement to Dexion Capital, an alternatives investment bank, reading:

> The insurance loss for the aircraft hull is currently estimated to be $100m. Liability payments for passengers could reach up to $500m depending on the nationality and actual cause of the incident. As such, total insurance losses will likely be between $500m and $600m.

The insurer of Air France Flight 447's Airbus A330 reportedly paid out 750 million dollars in 2009. Aviation insurance policies are legal contracts and have detailed clauses about indemnifying and compensating the families of aircraft accident victims. The most important of these clauses includes the need for there to be a proof of loss. In order to obtain this proof of loss—what Hugh Dunleavy was referring to—it is necessary for authorities to officially declare the aircraft lost. Until that happens, an insurance company is not contractually required to provide liability payments.

Hull loss insurance is usually straightforward. It's more confusing in the case of MH370 because it's unclear where the aircraft's hull is. It's hard to determine if the plane has been damaged beyond repair if there's no plane to look at. Hull insurance can, however, be used without seeing an aircraft according to a part of ICAO's *Standards and Recommended Practices* publication's definition of an accident, listed under its Annex 13 clause. In this, the definition has three basic components: 1. That a person is seriously or fatally injured as a result of the aircraft; 2. That the aircraft sustains extreme structural damage or failure; and 3. That the aircraft is missing or is completely inaccessible.

Awkwardly, this definition of an accident does not match the one in the US government's *Code of Federal Regulations*, the government's annual publication codifying the rules and regulations of the executive departments and agencies within the federal government, which matches that of Malaysia's Department of Civil Aviation and of most other civil aviation authorities.

The listing in the US government's Code of Federal Regulations (Title 49 → Subtitle B → Chapter VIII → Part 830 → Subpart A → §830.2) defines an accident as "an occurrence associated with the operation of an aircraft which takes place between the time any person boards the aircraft with the intention of flight and all such persons have disembarked, and in which any person suffers death or serious injury, or in which the aircraft receives substantial damage." Under this definition, the search for MH370 is not an accident investigation.

The discrepancy in these definitions of an accident may seem insignificant to most people. For the officials trying to figure out compensation for MH370, however, having two distinctly different definitions muddies the water.

Beyond insurance, when an event is classified by authorities as an accident, priorities immediately change in a dramatic way. Additional personnel are deployed, budgets are enhanced, and compensatory claims are settled. Typically, the investigator in charge (IIC), assigned by the responsible civil aviation authority, calls all the shots. However, until the aircraft is found and it is determined that all survivors have been cared for, the IIC will only assist the search and rescue teams by assuring complete access to the accident site and providing resources if needed.

Search and rescue operations (SAROPS) funds come from the government of the country of register of the aircraft in question—in this case, Malaysia— and of the country thought to be closest to the accident site—in this case, Australia. For MH370, other countries have offered to contribute to the search costs, but the Australian government hasn't accepted. The Australian Parliament's Treasurer, Joe Hockey, told reporters, "It is understood that the plane went down in waters that are our responsibility, and there is a cost to having responsibility. And we don't shirk that. We accept our responsibility, and we'll pay for it."

16

TIMELINE

When lacking physical evidence, safety analysts and investigators rely on the art of speculation and scenario building—we're well-versed in it. In the case of MH370, without scenarios and some level of speculation, no one would have any idea where to look for the aircraft. For investigators, this is a natural method— the families of passengers, crew, and other observers of this process, however, often find it cruel.

Suffering through what-if and perhaps-it-happened-this-way theories are painful. I understand that. Unfortunately, without physical evidence or a crash site, it is the only option available to determine where a missing aircraft might be and if the root cause of the accident has been mitigated. Until the aircraft is located, the root cause cannot be determined.

In some ways, this type of speculation can identify future vulnerabilities and risks more thoroughly than would an accident with a clear-and-present root cause. With that in mind, we begin with the few clues we have available to us. So what is a clue? *Webster's Dictionary Online* defines it as "something that helps a person find something, understand something, or solve a mystery or puzzle." Finding MH370 fits that description quite well, so what are the clues? The timeline for one:

Timeline:

1:21 a.m.: MH370's transponder failed to respond to secondary radar. The transponder failed to respond just after Kuala Lumpur ATC released the aircraft and before Ho Chi Minh ATC received it.

1:22 a.m.: MH370 disappears from Thai military radar.

1:22–1:28 a.m.: The aircraft changes course from Beijing toward Penang, Malaysia, where there is a MAS maintenance base. The crew fails to respond to multiple radio contact attempts from the tower and other nearby aircraft.

1:28 a.m.: A Thai radar station in southern Thailand's Surat Thani province picks up an unknown aircraft flying in a direction opposite to where Flight 370 had been traveling. The Thai data corroborates what the Malaysian military found earlier—that the plane did indeed turn around toward the Strait of Malacca. "The unknown aircraft's signal was sending out intermittently, on and off, and on and off," a spokesman for the Thai military told CNN on March 18, 2014. He said the Thai military lost the unknown aircraft's signal because of the limits of its military radar.

1:30 a.m.: Civilian radar loses contact with the plane. Malaysian air traffic controllers in Subang, outside Kuala Lumpur, lose contact with the plane over the Gulf of Thailand between Malaysia and Vietnam at coordinates N 06° 55.15' and E 103° 34.43'.

1:37 a.m.: Expected ACARS transmission doesn't happen. The ACARS signal was supposed to transmit a half-hour after it last did so, which would mean it should have sent a signal at 1:37 a.m., but it didn't. This means that the ACARS stopped communicating sometime between 1:07 and 1:37 a.m. It's a significant event, particularly because turning off ACARS, if that's what happened, is not simple; it requires a certain know-how. If the flight was hijacked, disabling ACARS would have been a strategic move.

2:15 a.m.: Malaysian military radar detects what's believed to be the plane. What they pick up is hundreds of miles off course of where MH370 should be, about 200 miles northwest of Penang.

2:22 a.m.: The Royal Malaysian Air Force detects the flight on its radar for the last time. The plane has swerved far off course, over the Malaysian coastal area of Penang in the direction of the Malacca Strait.

7:13 a.m.: Malaysia Airlines tries to make a voice call to MH370, but there is no pickup. (According to Malaysian officials in Beijing, there was no attempt at communication between Malaysia Airlines and MH370 for the five-hour period before this.)

8:11 a.m.: The sixth and final full satellite signal from the plane is sent.

Evidence released on March 25, 2014, shows there may have been a partial signal sent at 8:19 a.m. At 9:15 a.m., when ATC tries to send its next message, nothing goes through. The plane is no longer logged onto the network.

17

VALIDATED ASSUMPTIONS

In order to use the clues available to us to determine what happened to MH370, it is necessary to frame the scenarios we create by using facts or validated assumptions. The framing process grounds the scenarios. Without framing them, we could end up chasing rabbits into black holes and never reach any definitive conclusions. We need to have the freedom to imagine and play out how different possibilities could have panned out, but we also need to do this smartly and not get lost in Wonderland.

Below is a list I made of validated assumptions. I verified them with several experts, including two commercial Boeing 777 captains and a commercial aircraft maintenance quality control specialist who works for a number of airlines, including Malaysia Airlines. One of the captains I spoke with routinely flies from Kuala Lumpur to Ho Chi Minh on the same route MH370 began on March 7, 2014.

Validated Assumptions:

1. The pilot was not under duress or heightened awareness at the time of his last radio communication with Kuala Lumpur Area Radar at 1:21 a.m.

2. MH370's transponder was disabled at 1:21 a.m. Its second transponder was not enabled.

3. Malaysia Airlines has a good safety record (or did at the time of the plane's disappearance). It maintains its aircraft according to all regulatory standards, including the FAA's Airworthiness Directives, and regulations to correct unsafe products.

4. If the aircraft suffered from a mechanical emergency when it went off the radar, standard procedure would be for its pilots to fly to the nearest airport. The nearest airport known to the pilots was in Penang, Malaysia, and the second closet was in Langkawi, Malaysia.

5. The primary radar in the vicinity of where MH370 was at the time of its disappearance does not have the capability of determining altitude.

6. There is no technology that can see through water and definitively determine what metals are in the

ocean, as GeoResonance claims its technology can do.

7. No part of MH370's Boeing 777-200ER was modified outside the Type Certificate Data Sheet, the FAA's definitions of what defines well-made aviation devices.

8. There were two pilots in the cockpit during the transition between Kuala Lumpur area radar and Ho Chi Minh area radar. It's not likely that one of the pilots would have exited the cockpit during this transitional period.

9. Two passengers were using stolen passports.

10. It is possible to access the electronics and equipment (E & E) bay from the cabin area.

After examining the clues and validating these assumptions, I believe there are categorically only two explanations for the disappearance of MH370. Either the aircraft was commandeered by one of the pilots or an assailant or there was a fire in the cockpit or the equipment and electronics compartment on the left side of the aircraft. A fire would have put a severe pressure on the aircraft's electrical system and caused the electronics and surrounding insulation to produce a dense smoke, disabling communications and resulting in the depressurization of the aircraft.

18

THE HAND-OFF PROBABILITY

One of things that bothers me most about MH370 is the ever-present question of why the aircraft disappeared when it did—during the hand-off from Kuala Lumpur radar to Ho Chi Minh radar. If someone intended to make an airplane disappear, the ideal time to do it would be during the hand-off between air traffic control centers. During the hand-off, controllers expect a five- to ten-minute silence while radio frequencies are changed, checklists are performed, and pilots discuss amongst themselves who is flying the plane and who is handling the radios. It is also when the controllers releasing the aircraft divert their focus away from it on their radar screens and the receiving radar is not yet in direct communication with the aircraft. This means it is when the aircraft is most off the radar. It's the only time during the flight when ATC isn't watching it.

It seems highly unlikely that it's a coincidence that MH370's transponder failed and all communication with the aircraft ceased during the most vulnerable segment of the flight. Mechanical failures typically occur during a phase of flight when operational activity is at its peak, for example, during takeoff or landing, not the cruise portion of the flight. Takeoff and landing are when the mechanical components on the aircraft are being used the most, and therefore, statistically it is the most likely time for them to fail.

The flight from Kuala Lumpur to Beijing is approximately six hours, or 360 minutes. If all ten-minute periods are equal, the probability of an event occurring during any ten-minute period would be about 10/360, or 2.78 percent. But they're not equal. In calculating the likelihood of an event occurring during a particular time period, it is important to consider the number of events and the energy they require that occur during that time period and compare those numbers to the numbers of other time periods. If a device is used more extensively and requires more energy during one period than other periods, the likelihood of failure is greater during the period in question.

Calculating these probabilities can be particularly complicated because we must first count the number of mechanical or electrical events occurring during each phase of flight and know the energy that they require. There are an extraordinary number of operational events that require more-than-usual energy expenditures during a flight's entrance into and out of air traffic control centers; among these are changing radio frequencies, engaging autopilots, entering new frequencies into two different radios, and then pushing the swap button, which engages three mechanical relays, etc. Because of this, the probability of a mechanical or electrical failure during that ten-minute hand-off period is exponentially greater than the probability of one during any other ten-minute period of the flight.

19

THE INSIDIOUS NATURE OF HYPOXIA

Hypoxia can result from the effects of a mechanical or electrical failure, and the possibility that hypoxia played a role in MH370's downfall has been discussed in great detail. Hypoxia is extremely dangerous and has contributed to aviation accidents before.

Hypoxia occurs suddenly following a rapid or explosive decompression of an aircraft, and its onset is slow and undetectable. The higher the altitude, the less oxygen content is in the air and available to breathe. Hypoxia can be a silent killer and sometimes causes a sense of euphoria. It also causes light-headedness and a severe degradation of situational awareness, leading to potentially deadly decisions and a failure to react to emergency situations. If supplemental oxygen is not available, then loss of useful consciousness and asphyxiation follow.

The effects of hypoxia vary in severity. For example, my home in Colorado is 8,900 feet above sea level. My brother-in-law lives at sea level in Florida. If he travels directly from his home to my home, he develops altitude sickness, which causes headaches, nausea, and fatigue. At altitudes greater than 12,000 feet, the percentage of available oxygen in the air drops to levels that begin to affect awareness and perception. Above 22,000 feet, useful consciousness is not sustainable for more than ten minutes.

In the case of Helios Airways Flight 522, the aircraft crashed into the mountains near Marathon, Greece, on August 14, 2005. The passengers also suffered from hypoxia. After testing and repairing the air conditioning system of its Boeing 737, the pressurization mode switch was left in the manual mode. When the pressurization system is left in manual mode, the pilot must select the cabin altitude. In this case, the pilot did not realize the switch was in manual even though verifying the position of the switch is part of the preflight checklist.

Because the switch was left in manual, the outflow valve, which controls the pressure in the cabin by restricting airflow, remained open, allowing the pressurized air to escape. As a result, the cabin altitude climbed at the same rate as the aircraft, exposing the passengers and crew to extremely high altitudes and a severe deprivation of oxygen.

The cabin altitude warning sounded five minutes after takeoff, when the aircraft passed 12,040 feet. Two minutes later, the pilot contacted air traffic control and reported air conditioning problems. Six minutes later, the controller lost contact with the crew. Only eight minutes after the cabin altitude warning sounded, the crew was incapacitated from hypoxia at 28,900 feet. The autopilot stabilized the aircraft at 34,000 feet at 09:23 a.m., eleven minutes after the cabin altitude warning. Fighter jets were scrambled, and the pilot could be seen slumped over the controls.

At this point, the story of Helios Flight 522 takes on a whole new meaning and provides a vital clue in solving the mystery of MH370. At 11:49 a.m., two hours and thirty minutes after the crew succumbed to hypoxia, the fighter jet pilots saw a man in the cockpit trying to gain control of the aircraft. Flight attendant Andreas Prodromou, who held a UK commercial pilot license, had taken the pilot's seat and was attempting to descend to a safe altitude; unfortunately, the left and then the right engine flamed out from fuel exhaustion, and the aircraft crashed. The accident report stated that Prodromou remained conscious by using passenger air masks to get to the front of the plane, where he used a portable medical oxygen supply to gain access to the cockpit.

Hypoxia is a serious threat to aviation safety, yet little has been done to mitigate it. It never ceases to amaze me that the lessons learned from one aircraft accident appear over and over in subsequent accidents.

Accidents caused by hypoxia can be avoided by early detection and evasive action. Because hypoxia can effect a pilot without his or her knowledge, a system must be devised to assure the pilot is aware at an early stage. I recommend a simple solution to this problem: for transmissions to ATC to include cabin altitude in addition to aircraft altitude. This way, ATC would be able to identify if cabin altitude has exceeded 12,000 feet. If it's reported to ATC that it has, ATC would know to demand that the pilot make an emergency descent before his or her judgment becomes impaired. This simple upgrade could potentially save hundreds of lives in the next few years and would have answered many questions about the fate of the people on board MH370.

20

HATCHING A NEW THEORY
ON *PIERS MORGAN LIVE*

On March 14, 2014, one week following the disappearance of MH370, I was on *Piers Morgan Live* discussing what could have happened to the aircraft with other aviation accident analysts and Piers. Like the rest of the world, I had been searching for an answer to what had already become a mystery beyond compare.

Just prior to the show, as I was waiting in the green room, I received a message from the aircraft quality control specialist I know who has inspected Malaysia Airlines facilities. He and I had discussed the possibility of a hijacker stowaway waiting in the plane's electronics and equipment bay for the right opportunity to commandeer the aircraft. Considering the difficulty of getting to the E & E bay, we had considered this unlikely, but in our back and forth messages, he reminded me of an access point I hadn't considered.

The access point is behind the aircraft's forward galley, which is behind the pilot—just in front of the cabin door, there is a floor hatch providing access to the area. Once an E & E compartment is breached, an aircraft would be vulnerable to attack. A hijacker could disarm the very communication and navigations components that became unresponsive during MH370's journey.

Prior to the second block of the show, I spoke to one of *Piers Morgan Live*'s Producers, Winnie Dunbar, about this new consideration of mine, and Winnie asked if I was ready to talk about it on air. As I began to answer, she whisked me off to the set next to Mark Weiss, a former Captain of an American Airlines Boeing 777 and a security consultant for The Spectrum Group.

As Piers was heading to the a commercial break, I presented the possibility to him and Mark. The possibility took the emphasis off the cockpit being the place most likely to have been compromised and placed it on the E & E bay. When the second block began, Piers asked me to elaborate, and I explained that the Boeing 777 has a hatch that can be opened to allow a person to crawl down into the aircraft's equipment area. Once there, the person would have access to all the circuit breakers for the plane's equipment, including the ones responsible for the transponder and ACARS.

I had not discussed bringing up this possibility before the show, so when Piers heard it, he was taken aback. He asked Mark what he thought about it, and without hesitation, Mark agreed that it was a good possibility to consider. He expressed his opinion that if the plane had been compromised through this, two people would have been needed: one to mess with the equipment and one to fly the aircraft.

For many weeks after the appearance, I was inundated with requests from around the world for appearances on other shows and for radio interviews. Ten days after the episode, a Boeing spokesperson, Gayla Keller, said in a statement to WTOP radio in Washington, D.C., that Boeing wouldn't address the idea of a hijacker gaining access to the aircraft. "We won't engage in speculation regarding this," she said.

21

THE NEW SEARCH

The search for MH370 was temporarily put on hold on May 26, 2014, when Martin Dolan, Chief Commissioner of the Australian Transportation Safety Bureau, made an announcement that ATSB would be looking for a company to take over the search.

The first search that ATSB conducted was severely hampered because it didn't have a bathymetric map, an underwater topographic map, of the search area. Without this, its search vessels had been forced to travel at painfully slow speeds to avoid dragging the valuable towed pinger locater into the side of an underwater mountain. The lack of a map of the ocean floor also impacted the ability of the autonomous underwater vehicle. It's important for an AUV to maintain a stable height above the ocean floor to obtain good sonar images. With a detailed map of the ocean floor, the track of the AUV could have been pre-programmed. This would have allowed the vehicle to anticipate steep inclines and deep crevasses, providing for much higher speeds while searching.

In June 2014, ATSB signed a three-month lease for a commercial survey ship, *MV Fugro Equator*, to map the ocean floor in an area five hundred miles southwest of the area that was searched in April and early May. The new search area is an arc 960 miles west of Perth, Australia, and is 150,000 square kilometers, about the size of Florida. The new search plan does not include an AUV. The Dutch company that owns and operates the *Fugro* is using a towed sled unit that is permanently connected to the ship by a cable containing power leads and glass fiber data communications lines. The information retrieved from the sled unit is shipped to shore in real time and is ready for evaluation the next day. This is a sharp contrast to the days and sometimes weeks it took during the first phase of the search to send the AUVs to shore and analyze the data they received.

22

LESSONS LEARNED?

I wish I could tell you that what happened to MH370 was a fluke, that it definitely never happened before, and is unlikely to happen again. But I can't. MH370 is unique in its story because of how long it's taking to find the aircraft, but I have written about similar situations before. Countless recommendations and proposals that would improve safety have failed to catalyze into results. Recommendations in this chapter will likely be added to the list of previous attempts by me and other safety professionals to prevent loss of life that have gone overlooked or put on hold.

Thirty-one years ago, I worked for Roy Morgan, who owned Air Methods Inc., an emergency medical helicopter company. Roy is brilliant and started the company to help save lives. On road trips we took to sell our services to hospitals across the country, Roy told me fascinating stories of times spent flying helicopters in the mountains of Colorado while patrolling power lines for a public service company.

Roy was often asked to assist in search and rescue operations for stranded hikers and climbers in the wilderness. There was one particular search and rescue operation he'd embarked on that shook him to his core, changing his life. The story he told was about his attempt to help a young mountain climber who was badly injured from a fall. The climber had a radio with him and was able to communicate with Roy, and as Roy flew his helicopter back and forth desperately trying to locate the injured climber, he talked with the young man. The man could hear the helicopter and tried to guide Roy to his location, but the echoes off the Rockies made that impossible. Ultimately, Roy had to return to the airport to refuel and was not able to raise the young man on the radio again. A week later, rescuers found him deceased at the foot of a cliff with his radio in his hand.

Roy approached me one day and asked me to join him on a visit to see a friend of his who claimed to have an invention that could help with search and rescue operations. Little did I know that visit to a kind, quiet man in Denver would be so incredibly impactful. The man answered the door with a huge grin and welcomed us in. To be cordial, he offered us some coffee, but I could tell he was incredibly anxious to show us his invention, so we passed and followed him into his basement. The basement was a bit of a hoarder's paradise, and the low ceilings gave the room an eerie vibe. He told me his name was Richard, "but my friends call me Ricky," he said.

Ricky asked if I could help him demonstrate how the device worked, and he handed me a classic 1970s Spiderman walkie-talkie. He asked me to stand back a few feet, so I stepped back while he pulled out what appeared to be a pie tin—not a cheap aluminum one, a real pie tin. He held it upside down, and I could see twelve small bolts sticking through the sloped sides. Each of the bolts was wrapped with thin wire. Some wires connected back to a small Heathkit project box with what looked like a compass attached to the top of it.

Ricky asked me to press the walkie-talkie transmit button, and when I did, the needle on the compass spun and pointed directly at me. Roy smiled at me from across the room. "What do you think? Neat, huh?"

I could see the excitement on Ricky's face when he saw my amazement as the needle attached to the seventy-five-cent pie tin followed me and my Spiderman walkie-talkie around the basement. It was all too simple, too easy to be true. I asked how in the world he came up with the idea. He smiled through his overgrown beard and said, "It ain't rocket surgery. You just gotta figure it out and do it, that's all."

Roy paid Ricky a couple hundred bucks, and as he packed the invention into a box, he told me how we could install them on the six emergency medical helicopters we operated—we'd just need a bolt here and a nut there. As it turned out, we did install them. In fact, we installed a lot of them. Although they were no longer made of pie tins, we manufactured dozens of them because they worked. The search and rescue teams even liked them and tried them for a while, but they were more effective in the air.

We built the contraptions for seventy-five dollars in materials and about eight hours of labor. The FAA and FCC approvals took about a year, but we were able to install them in the meantime, and they saved lives. They continue to save lives in one form or another today.

It boggles my mind that thirty-one years ago, we could find a two-dollar Spiderman walkie-talkie twenty miles out in the woods, but today we don't have the technology to find a 350 million dollar, 600,000-pound airliner with 239 people on board.

23

THE SANDPILE EFFECT

Without question, the complexity of air travel has reached a level that Orville and Wilbur Wright could never have conceived. An aircraft's systems have become almost entirely interdependent. Nearly every flight function is monitored and evaluated by a centralized computer, and there are as many as five separate synchronized computers. When anomalies or negative interactions are brought to light, engineers are faced with developing a fix to mitigate the problem. In this highly advanced technological world, this is a daunting task. Because everything is connected and dependent on everything else, making changes of any kind has the potential to create a dramatic adverse effect.

Here's an example: let's say there's an aircraft model—we'll call it model A—that was manufactured with electronic fuel controls for its engines, and a pilot of one of the planes was making an approach into a high altitude airport. When the aircraft neared the ground, a wind shear affected the landing, so the pilot elected to make a go-around and try the approach again. This time around, it took too long for the throttles to come forward, so the needed thrust was not available to make a successful go-around, and the plane crashed off the end of the runway. (None of the imaginary passengers were injured.)

A slight change in the software could have corrected the problem and given the pilot the ability to override the spool time in the event of an emergency. This could theoretically damage the plane's engines, but at least the aircraft wouldn't crash.

If you were the plane's engineer, would you have installed this type of software? It sounds like the right thing to do, doesn't it? Here's the problem: The spool time is limited because of the vast amount of torque generated by the plane's engines as they spool up. By increasing the spool speed, you would dramatically increase the torque on the engine mounting bolts, the wing attachments, and possibly even the rudder of the airplane. These stresses would not present themselves until the additional torque loads manifest into a critical failure. Remember, every piece of equipment on the airplane is linked with the others, from the coffee pot to the radios to the tires. They all play a vital role in a successful flight.

Like aircraft, the safety management systems in aviation have evolved to the point where all their components are also interdependent. A small change in the way a hazard is identified in an airline can have a dramatic effect on an airline's ability to plan for or react to potential risks. The safety system has become just as complex as the operational systems it protects.

One way to understand this is to compare it to a sandpile. A sandpile is created by dropping grains of sand on top of previous ones. As each grain of sand is dropped upon the pile, the pile gets bigger and bigger. Then, suddenly, when a particular grain of sand is added, an avalanche of sand rushes down the sides of the pile and keeps running until everything is balanced again. This is just like an aircraft. Its system can get bigger and bigger as it gets more and more modern, but a bad connection with just one system can affect all of the others. It can bring an avalanche of operational failures.

Systemic failures are even more worrisome because one avalanche-causing systemic failure could be extremely far-reaching, affecting more than one aircraft. It could even be catastrophic. For this reason, airline executives and civil aviation authorities take great care before mandating even the slightest of safety improvements. They need to avoid upsetting the delicate balance of such tightly-woven safety systems.

In order to effectively manage new processes and procedures in the FAA, we utilized a systems architecture model. As the Information Technology and Business Process Lead for FAA flight standards, I was responsible for assuring the business processes and changes to them were fully supported by information technology systems.

Our systems architecture model was like a conceptual wiring harness that organized the structure and behavior of our safety system and the relationship amongst its parts. When an airline made a request to add a new route to its flight plans or a new aircraft to its fleet, our system architecture model could be used to identify which safety processes might be affected and could mitigate associated risks prior to approving the request. The architecture model filled the walls of a large conference room, but it allowed me to literally walk through proposed changes and system improvements before they were implemented. I was able to drop the next grain of sand onto the sandpile and see if it would cause an avalanche.

24

CONCLUSIONS

Making conclusive statements without a preponderance of evidence is a fool's errand. That being said, I compiled what I believe to be the only credible information about the disappearance of MH370, made what I believe to be reasonable assumptions in order to provide scope, and input the assumptions into a Bayesian model I created in which I could play out what a scenario would look like if it had happened.

Here are what I consider the most likely to least likely events that led to the plane's downfall.

1. Electrical outage
2. Sabotage
3. Hijacking/Terrorism
4. Pilot suicide
5. Wireless control

I believe that what most likely happened to MH370 is that Pilot Zaharie Ahmad Shah detected an electrical outage followed immediately by severe smoke in the cockpit less than one minute after signing off with Kuala Lumpur ATC. The smoke in the cockpit was severe enough to start a fire in the cockpit, so he followed the checklist protocol in the event of this type of emergency. First, the oxygen masks, then load shedding, and then pulling circuit breakers to mitigate the possibility of an electrical short that may have initiated or may be contributing to the fire.

Pilot oxygen masks are designed to provide an airtight seal around the mouth, nose, and eyes. The mask is equipped with a built-in microphone for communicating with air traffic control and making announcements to the cabin. What I believe happened on MH370 is that because of the airtight seal, sound could not escape the mask after it had been donned, and the Captain therefore was not able to communicate with the First Officer or ATC.

Realizing the severity of the situation, he elected to make an immediate turn toward the closest airport with adequate runway length to accommodate the landing. This would have been either Langkawi or Penang, where Malaysia Airlines has a maintenance base.

The load shedding was unsuccessful in reducing the smoke in the cockpit, although it was successful at shedding the communications bus, including the transponder. In a desperate effort to quell the smoke, the pilot depressurized the aircraft. The lack of oxygen at 35,000 feet would most likely extinguish the fire. At this point, the oxygen masks dropped in the cabin for the passengers. The smoke in the cockpit continued to build. And the fire, likely in the E & E compartment, is intense. The recently replaced metal-cased oxygen lines for the pilot's oxygen had been replaced with plastic lines in accordance with an Airworthiness Directive, but those lines are not resistant to high temperature and melt under extreme heat, depleting the cockpit crew's oxygen supply within seconds.

There is, of course, no way to know definitively what may have caused a fire or intense smoke. However, a fire in the E & E compartment would explain the loss of the transponders and communications equipment, as all of the communications radios are on the left side of an E & E compartment. Extreme heat in this area could have caused a chain of events that compounded to cause the failure of all communications, temporary incapacitation of the crew, and the asphyxiation and death of the passengers. The opposite side of the E & E compartment houses the autopilot gyros and equipment. A fire on the left side of the E & E compartment would have to travel more than fifteen feet to affect the autopilot.

In the Boeing 777-200ER, there are at least two medical oxygen bottles on board. It's feasible that either one of the pilots or one of the flight attendants obtained a medical oxygen bottle and was able to survive long enough to wait out the fire in the E & E compartment and the smoke in the cockpit. This would be similar to how one of the Helios Flight 522 flight attendants was able to survive for two and a half hours using a medical oxygen bottle at 34,000 feet before the aircraft ran out of fuel and crashed.

Malaysia Airlines' flight attendants are trained in the use of the autopilot. In cases of extreme emergency, they would be capable of flying the aircraft using the autopilot system. If a flight attendant was able to get into the cockpit and use the autopilot, he or she would not have been able to contact ATC and would most likely have had no idea what the plane's location was. The person would not have been able to find an airport or land the plane safely. Knowing the worst was inevitable, the best option would have been to avoid massive fatalities on the ground by avoiding city lights and flying over the ocean, navigating around Indonesia and into the Indian Ocean.

25

FROM TRAGEDY TO CALAMITY, AND WHERE WE GO FROM HERE

In public relations classes in Palm Coast, Florida, at the Federal Aviation Administration Center for Management and Executive Leadership, we were taught to use the word tragedy when discussing fatal aircraft accidents. To call MH370's disappearance a tragedy seems an understatement. The events immediately following the aircraft's disappearance and many thereafter would be better described as calamity.

All the ground—or ocean—investigators are painstakingly looking for the aircraft. Remote investigators, such as myself, engineers, and journalists are digging deep to find clues that could help and hopefully ease the pain for the families. We may never fully agree about what happened moments after Captain Zaharie signed off with Kuala Lumpur radar. Whether

the plane was commandeered by a terrorist or someone on board, crippled by a fire that poured disabling
smoke into the cockpit, or suffered a catastrophic loss
of all electrical systems, something happened that
caused MH370 to vanish.

Regrettably, it is impossible to implement a plan to
prevent the recurrence of a root-cause without tangible
evidence of what the root-cause was. Even hard evidence
with thorough analysis and detailed recommendations
following the Air France Flight 447 investigation
resulted in little or no substantive safety improvements.
Influencing change without conclusive proof stands little
to no chance of bringing results. However, the apathy
displayed in the hours following the disappearance
of MH370 implies serious systemic flaws that can be
thoroughly analyzed and improved upon.

The lack of swift and definitive action by area radar
control, the misleading information from Malaysia
Airlines, the lack of communication between civilian
and military radar operators, and the inability to provide the victims' families with full reports are all factors that elevated MH370 from a tragedy to a calamity.
These systemic failures need to be fixed, and their fixes
are not difficult ones to implement into standard protocol. There are lessons to be taken from MH370 that
will prevent future calamities.

I once again join the plea for required in-flight
tracking of aircraft, real-time streaming of safety

and location data, emergency locator transmitters that actually locate ditched aircraft within minutes so rescue of survivors may be possible, and batteries on underwater locator beacons that ping for a long enough time and broadcast at a lower frequency, tripling the detection range to actually find the black boxes.

I would also like to add the following safety recommendations, which would have an even higher probability of preventing another accident before it happens:

1. Require ULBs to have unique identifying pulse patterns, such as a Morse code that transmits the serial number of the unit, so searchers can distinguish the signal from other oceanic devices.

2. Require both transponders, ADS-B and ADS–C, to transmit cabin altitude to ATC, who could immediately invoke the Mayday handling procedure for aircraft that transmit cabin altitudes above 12,000 feet from Mean Sea Level (MSL).

I may be chasing windmills and hoping for the impossible. When I hear stories from regulators about how long it will take and how much it will cost to implement, I remember Ricky and his pan, and I think, "Just figure it out and do it!" There's a whole world out there to explore, and everyone should safely be able to do so. Their planes should never disappear.

26

THE ACCIDENT ALGORITHM

The Bayesian formula I used to determine the most likely scenarios that led to the downfall of MH370 was computed through a logic program I developed and call the "accident algorithm." I've been using it in my investigations for more than twenty years, and over and over again it's accurately determined what caused an accident.

So, how does it work? Every accident leaves a series of clues, or what seem like clues, and dozens of pieces of information can be extracted from them. I work with eight experts to extract the information and put it into a matrix I created that processes the formula. The information is weighted by a confidence factor that considers the source of the information as well as its logical viability as determined by the experts. The experts then compare each piece of information against each of the others—assessing interrelationships—and provide a numerical score to each interrelationship. The numerical scores are between negative five and five. A negative five means the expert has compared the information against the other and they completely contradict one another. A five means the expert has compared the information against the other and they are completely complementary and support the argument that the information is true. A zero means the expert compared the information against the other and they either have no supporting or complementing relationship or it is outside the realm of the expert's experience and knowledge.

Through a series of computations, the facts that the experts have the most confidence in rise to the top of the matrix and show, with all the experts' opinions factored in, which scenarios are most likely to have occurred. The interrelationships between the most likely pieces of information can then be considered together and put together a story showing what most likely happened and why.

The information being considered to establish what likely happened to an aircraft falls into the ten areas below. With different subject areas, different clues, and different extracted information, the same formula could be used to solve other mysteries.

Subject areas to determine the likely cause of a plane crash:

1. Systems safety of the aircraft

2. Operations of the pilot

3. Equipment maintenance

4. Air traffic control

5. Avionics systems and radar

6. Political aspects of a region

7. International security

8. Legal issues surrounding the companies involved and the passengers

9. Regulatory factors regarding the plane

10. The qualifications of everyone associated with the flight.

In total, I put 128 pieces of information into my formula to determine what most likely happened to MH370. I have included a number of them below, as well as the sources of many of them. Most of sources I've listed are from CNN but the information can be found elsewhere too. If no date is given to correspond with a time, the corresponding event happened on the day that MH370 went missing, March 8, 2014.

The matrix showing the expert evaluations and results, as well as further information regarding the interpretation of the data can be found on my website, www.whyplanescrash.com.

Input information:

1. March 8, 2014: Flight 370, a Boeing 777-200ER, disappears on what should have been a routine flight from Kuala Lumpur, Malaysia, to Beijing. There are 239 people aboard—227 passengers and twelve crew members.

2. 12:41 a.m.: MH370 takes off. All tracking systems are working as the aircraft leaves Kuala Lumpur. (CNN.com, March 23, 2014)

3. 12:42 a.m.: Flight 370 is cleared to climb. (ATC audio recording)

4. 1:07 a.m.: The last ACARS transmission is sent and shows a normal routing all the way to Beijing. (Press statement by the Malaysian Ministry of Transport on March 23, 2014, reproduced in a CNN.com article on March 23, 2014)

5. ACARS is the onboard computer that collects information—a lot of it—about aircraft and pilot performance. It's akin to the computers in automobiles that track oil levels and engine performance.

6. 1:19 a.m.: The captain radios his last words to the Kuala Lumpur ATC. (Attachment to a preliminary report by Malaysia's Transportation Ministry)

7. Controllers tell the MH370 captains to check in with their counterparts in Ho Chi Minh City, Vietnam.

8. Flight 370's pilot, Captain Zaharie Ahmad Shah, is the last person on board the jet to speak to air traffic controllers, telling them, "Good night, Malaysian three seven zero." There is nothing unusual about the captain's voice in this message; it conveys no sign of stress. (CNN.com referencing news from Malaysian authorities, April 11, 2014)

9. Police played the recording to five Malaysia Airlines pilots who knew MH370's pilot and copilot and were told by each that there were no third-party voices in the recording. (CNN.com referencing news from Malaysian authorities, April 11, 2014)

10. It's significant that there were no other voices heard in the recording because it suggests that the captain was working the radio while the first officer was flying the plane. (CNN.com, April 11, 2014)

11. 1:19 a.m.: Captain Zaharie Ahmad Shah speaks to Kuala Lumpur ATC and does not say that the plane has run into any trouble. (*Wall Street Journal*, May 2, 2014)

12. Flight 370's pilots do not radio Ho Chi Minh City, Vietnam, as they are supposed to. (*Wall Street Journal*, May 2, 2014)

13. Military radar tracks MH370 between 1:19 a.m. and 2:40 a.m. (CNN.com, March 23, 2014)

14. 1:21 a.m.: The aircraft's transponder stop communicating.(Azharuddin Abdul Rahman, director of the Malaysian Department of Civil Aviation, at a news conference on March 12, 2014)

15. Shutting off a transponder is a simple turn of a switch in the cockpit. If this happens, an air traffic controller should notice and it should cause alarm. (CNN aviation correspondent Richard Quest, March 23, 2014)

16. 1:21 a.m.: MH370's crew should have contacted Ho Chi Minh ATC but don't, and it isn't until seventeen minutes later that Ho Chi Minh ATC asks Malaysian ATC where the plane is.

17. It can be assumed that for those seventeen minutes, Kuala Lumpur ATC either didn't notice the plane had gone missing or didn't act (CNN aviation correspondent Richard Quest, May 1, 2014).

18. The final ATC (secondary) radar fix occurs at 1:22 a.m. (Australian report).

19. 1:21 a.m.: MH370 disappears from Malaysian ATC radar while near airspace managed by Vietnamese ATC. No distress call or information is relayed (Malaysia Airlines statement, March 10, 2014).

20. The transponder stops working between 1:21–1:28 a.m. and immediately after, the plane appears to have changed course, heading west.

21. A person "familiar with the investigation" says that, "It might be a coincidence, but if you are choosing the one moment in the flight to go dark, that's the moment. If it is just a mechanical failure, it is an extraordinary coincidence." (*Wall Street Journal*, May 2, 2014)

22. The Malaysian government has not said when, or if, the plane was reprogrammed to fly off-course. (CNN.com, March 23, 2014)

23. The aircraft deviated from the flight-planned route. (Australian report)

24. 1:22 a.m.: An unknown aircraft that Thai military radar has been tracking disappears from its radar. (Royal Thai Air Force spokesman tells CNN on March 18, 2014, CNN.com, March 23, 2014)

25. 1:28 a.m.: ATC in Thailand's southern Surat Thani province picks up an unknown aircraft flying in a direction opposite to what Flight 370 had been traveling. (Royal Thai Air Force spokesman tells CNN on March 18, 2014, CNN.com, March 23, 2014)

26. The Thai data corroborated what the Malaysian military found—that the plane turned around

toward the Strait of Malacca. But the radar contact was short-lived. A spokesman said the aircraft was only sending out signals intermittently, on and off. (CNN.com, March 18, 2014)

27. About 1:30 a.m.: Thai ATC radar loses contact with MH370.

28. Malaysian ATC in Subang, outside Kuala Lumpur, loses contact with the plane over the Gulf of Thailand between Malaysia and Vietnam at coordinates 06°55'15"N and 103°34'43"E. (Malaysia Airlines CEO, Ahmad Jauhari Yahya on March 13, 2014, CNN.com, March 23, 2014)

29. Radar is not reliable over the ocean or, as may have been the case with MH370, at low altitudes. (CNN)

30. 1:37 a.m.: Expected ACARS transmission, a half-hour after the last one, doesn't happen, which means that ACARS stopped communicating sometime between 1:07 and 1:37 a.m. (Malaysia Airlines CEO, March 13, 2014)

31. It's particularly significant that the transmission didn't go through because turning off ACARS takes know-how. If the flight was hijacked or a target of terrorism, cutting off ACARS would have been strategic because the system reports to satellites anything being done to the aircraft. (CNN aviation correspondent Richard Quest, CNN.com, March 23, 2014)

32. 1:38 a.m.: Ho Chi Minh ATC contact Kuala Lumpur to let the controllers know that it has not heard a word from the plane. "Verbal contact was not established," the transcript said. The two control centers then begin a conversation about communication attempts with Flight 370 and previous radar blips along its path. They speak every few minutes. (CNN.com, April 30, 2014)

33. 2:03 a.m.: The operational dispatch center of Malaysia Airlines sends a message to the cockpit instructing the pilot to contact ground control in Vietnam. The MH370 crew does not respond to the message. (Sayid Ruzaimi Syed Aris, an official with Malaysia's aviation authority, told CNN. com, April 29, 2014)

34. 2:03 a.m.: The first of two messages come from Malaysia Airlines that may have taken more precious time. The plane was in Cambodian airspace, the airline told Kuala Lumpur ATC. The Malaysians first pass the message on to Vietnamese ATC before trying to confirm Malaysia Airlines' news with Cambodian ATC. The airline later confirms its reassuring message. It has been able to "exchange signals with the flight," which was in Cambodian airspace, the transcript reads. (CNN.com, April 30, 2014)

35. But an hour after Flight 370 signed off, Vietnamese controllers poke holes in Malaysia

Airlines' message. The flight has not been scheduled to fly over Cambodia and officials there have no information about the plane nor have they had contact with it. (CNN.com, April 30, 2014)

36. 2:15 a.m.: Malaysia Airlines Operations Center tells Kuala Lumpur ATC that the plane is in "normal condition" and signals showed it was flying in Cambodian airspace. (*Wall Street Journal*, Malaysian Investigators' Report Timeline, May 1, 2014)

37. 2:15 a.m.: Malaysian military radar detects what's believed to be the plane. If it's the aircraft, it's hundreds of miles off course, about 200 miles northwest of Penang. (Malaysia's Air Force Chief, Rodzali Daud, in briefing on March 12, 2014)

38. According to a Malaysian Air Force official, military radar tracked the plane, even though it wasn't transmitting any information, as it passed over the small island of Pulau Perak in the Strait of Malacca. At that point, the plane was hundreds of miles off course, on the other side of the Malay Peninsula. Military radar showed that it then flew in a westerly direction back over the Malay Peninsula. It is then believed to have either turned northwest toward the Bay of Bengal or southwest, toward somewhere else in the Indian Ocean. This was the last time any civilian or

military radar is known to have tracked the aircraft. (Malaysian Prime Minister Najib Razak, and CNN.com, March 23, 2014)

39. The Malaysian military hands over its raw radar data to US and British officials, apparently setting aside concerns about any sensitive military intelligence. According to CNN aviation analyst this is a "huge" development in the case because "They don't want anyone to know how good their radar is . . . [but] obviously decided that doesn't matter." (CNN.com, March 23, 2014)

40. 2:22 a.m.: The Royal Malaysian Air Force pick up the flight for the last time on its radar system. It's believed that by this point, the plane has swerved so far off course that it is over the Malaysian coastal area of Penang, in the direction of the Malacca Strait. (Sayid Ruzaimi Syed Aris, CNN. com, April 29, 2014)

41. Data indicates that the aircraft flew an additional six hours after its last radar contact. (Australian report)

42. The Malaysian prime minister has said the military tracked the plane as it headed back across Malaysia.

43. A playback of a recording from military primary radar revealed that an aircraft that may have been Flight 370 had made a westerly turn, crossing

Peninsular Malaysia. The search area was then extended to the Strait of Malacca; but it's unclear when that happened. The report makes no mention of the military's role the night of the disappearance. (CNN.com, April 30, 2014)

44. Radar signatures offered evidence that the aircraft turned west after its last contact with ATC, and that contact was lost over the Strait of Malacca. But radar operators did not see it in real-time, meaning an opportunity to track the plane while it was in flight may have been lost. (CNN.com, April 1, 2014)

45. During the plane's last contact with ATC, as it was approaching Vietnamese airspace, it was reportedly flying at a cruising altitude of 35,000 feet. But military radar tracked it changing altitude after making a sharp turn as it headed toward the Strait of Malacca. (CNN.com)

46. According to an official who spoke with CNN, but was not authorized to speak with the media, the plane flew as low as 12,000 feet at some point before it disappeared from radar. The official said the area the plane flew in after the turn is a heavily traveled air corridor; that flying at 12,000 feet would have kept the jet well out of the way of that traffic; and that military radar tracked the flight between 1:19 a.m. and 2:40 a.m. the day it went missing. It was not clear, however, how long it

took the plane to descend to 12,000 feet. (CNN. com, March 23, 2014)

47. Prime Minister Razak said he believes there was someone monitoring the Malaysian military radar that picked up the flight, "but the interpretation was done after the event." He said it is not known whether the plane was MH370 and that no planes were sent up to investigate "because it was deemed not to be hostile [and] ... "behaved like a commercial airline, following a normal flight path." (CNN.com, April 24, 2014)

48. It's been said that the Boeing jetliner disappeared from military radar for about 120 nautical miles after it crossed back over the Malaysian Peninsula. This means the plane must have dropped to an altitude of 4,000–5,000 feet. (A senior Malaysian government official and a source involved in the investigation told CNN, CNN.com, April 11, 2014)

49. 2:22 a.m.: Military radar last records the plane as it is flying west past the Strait of Malacca and out into the ocean (according to a map provided with the investigator's report on May 1, 2014). Experts believe that the plane made a mystifying turn to the south shortly thereafter.

50. 2:35 a.m.: Members of Malaysia Airlines give another seemingly reassuring message. They say the airliner is "in normal condition based

on signal download." They place it off the coast of Vietnam and make it seem like the flight is back on track to its destination of Beijing. "We have two very unhelpful contributions from Malaysia Airlines—one suggesting the plane is in Cambodia, the other saying everything's normal. Neither is true," CNN aviation correspondent Richard Quest said. (CNN, April 30, 2014)

51. 2:35 a.m.: Ho Chi Minh ATC queries about status of MH370 and is informed that the watch supervisor is talking to the company. (Preliminary report released by Malaysian officials)

52. 2:35 a.m.: Malaysia Airlines' operations center tells the Kuala Lumpur ATC that a signal download from the plane indicates it is halfway up the Vietnam coast, on a regular flight path. (*New York Times* online, May 1, 2014)

53. According to Malaysia Airlines officials, Malaysian ATC told them at 2:40 a.m. that Flight 370 was missing from radar. (CNN.com, March 23, 2014)

54. 2:40–3:45 a.m.: Malaysia Airlines' preliminary search: Company officials said that during this time, the airline "sourced every communication possible to locate [MH370's] whereabouts before declaring that it had lost contact with the aircraft. [And] during this period of uncertainty, Malaysia

Airlines needed to establish facts by contacting other air traffic controllers and aircraft flying within the same route." (CNN.com, March 23, 2014)

55. 3:30 a.m.: Nearly an hour after the 2:35 a.m. message, Malaysia Airlines officials qualify their previous information. Their new message is that "The flight tracker information was based on flight projection and not reliable for aircraft positioning." (Malaysia Airlines' message transcript, CNN.com, April 30, 2014)

56. Malaysia Airlines releases a message at 3:30 a.m., but two more hours will pass before ATC notifies rescuers. In the meantime, controllers in Kuala Lumpur and Ho Chi Minh query each other and the airline. Kuala Lumpur ATC contacts its counterparts in Singapore, Hong Kong, and Beijing. (CNN.com, April 30, 2014)

57. The airline elaborated on its report, saying the "exchange of signals" it had referred to was actually the data from the automated flight tracking system, which didn't reflect changes in the aircraft's course. Airline officials concluded that Flight 370 was over Cambodia because a zoomed-in view of the flight tracking screen showed the text "Cambodia."

58. The airline appeared to deflect criticism for the actions it took the night MH370 disappeared,

saying in a statement that it's the job of air traffic controllers—not airlines—to keep track of planes in flight. (CNN.com, May 2, 2014)

59. 3:30 a.m.: Malaysia Airlines tells ATC that the information it had given was "not reliable for aircraft positioning." (Malaysian Investigator Timeline, May 1, 2014)

60. 3:45 a.m.: Malaysia Airlines issues alert. Its officials said it issued a "code red" alert that the plane was missing from radar, and that a "code red" is used when a crisis requires immediate deployment of emergency response plans. Officials said it took about an hour to issue the alert because the company was trying to confirm the plane was missing. They said that to verify this, they used various measures, including sending messages to the plane and awaiting a response. (CNN.com, March 23, 2014)

61. 5:20 a.m.: A Malaysian official says "MH370 never left Malaysian airspace." (CNN.com, April 30, 2014)

62. 5:30 a.m.: Malaysian ATC alerts a rescue coordination center that the plane is missing. (CNN.com, April 30, 2014)

63. It took ATC more than four hours after their last conversation with the pilots to activate rescuers to look for the missing plane. (CNN.com, May 2, 2014)

64. 5:30 a.m.: The watch supervisor at Kuala Lumpur ATC activates the Kuala Lumpur Aeronautical Rescue Coordination Centre (ARCC) (Malaysian Preliminary Report)

65. ARCC's search and rescue operations begin in the South China Sea, the site of the aircraft's last known position. (Malaysian Preliminary Report)

66. A recording from military primary radar reveals that an aircraft that might have been MH370 made an air-turn back onto a westerly course, crossing Peninsular Malaysia. The search area is then extended to the Straits of Malacca. (Malaysian Preliminary Report)

67. MH370 should have landed at the Beijing Capital International Airport at 6:30 a.m. on March 8, 2014. (CNN.com, April 29, 2014)

68. 6:51 a.m.: Ho Chi Minh's ATC makes a broadcast call on emergency frequencies to MH370 asking the crew to call them. (ABC News, March 15, 2014)

69. Sayid Ruzaimi Syed Aris said that at 7:13 a.m. Malaysia Airlines officials tried to "make a voice call to the aircraft, but [there was] no pickup." According to Malaysian officials in Beijing, there was no direct communication between Malaysia Airlines and MH370 for a five-hour period, until the airline tried unsuccessfully to call the cockpit. (CNN.com, April 29, 2014)

70. An Australian report says the plane received an unanswered ground-to-air telephone call at 7:13 a.m.

71. A public announcement of the plane's disappearance is made at 7:24 a.m. (CNN.com, March 23, 2014)

72. The sixth and final handshake (satellite transmission) from MH370 takes place at 8:11 a.m. According to Inmarsat data, transmission showed that Flight 370 was far south of where it should have been. (Prime Minister Razak at a news briefing on March 15, 2014)

73. The U.S. National Transportation Safety Board and the Federal Aviation Administration, along with Britain's Air Accidents Investigation Branch, concur with Malaysia's prime minister that the signal received on 8:11 a.m. was from Flight 370. (CNN.com, March 23, 2014)

74. Prime Minister Razak said that due to the type of satellite data received, they were unable to confirm the precise location of the plane when it last made contact. Authorities said they believe the plane was in one of two flight "corridors," either a northern route stretching to northern Thailand, Kazakhstan, and Turkmenistan in Central Asia, or a southern route toward Indonesia and the southern Indian Ocean. (CNN.com, March 23, 2014)

75. If the plane made its last satellite communication at 8:11 a.m., it continued to fly for seven hours after it first went off radar.

76. Malaysian officials told Chinese families that, by their calculations, the aircraft would have run out of fuel seven hours and thirty-one minutes into the flight. "Based on the fuel calculation ... the aircraft fuel starvation will occur at time 8:12," said Subas Chandran, a Malaysia Airlines representative. (CNN.com, April 29, 2014)

77. The last message received by the satellite ground station was at 8:19 a.m. (Malaysian Preliminary Report)

78. 10:30 a.m.: The full-blown search for the missing aircraft finally begins. At this point, it's been more than nine hours since the plane first stopped communicating.

79. 11:14 a.m.: Malaysia Airlines holds a news conference confirming the loss of contact with its aircraft. A spokesman said that their focus was "to work with the emergency responders and authorities, and mobilize its full support." The spokesman also confirms the nationalities of the people on board the aircraft. (NBC News, March 8, 2014)

80. In the news conference, the official cites "satellite information" that is causing Malaysian officials to

focus their attention on two massive "arcs" on both sides of the equator. (NBC News, March 8, 2014)

81. Because Malaysian officials believe that the plane could have taken one of two very different paths, their search area blankets 2.97 million sq. miles, nearly equivalent to the size of the continental United States. The northern arc stretches over Cambodia, Laos, China, Kazakhstan, and Turkmenistan. Officials later dismiss this arching route, saying no crashes or even a sighting of a wayward Boeing have been reported. (CNN.com, April 7, 2014)

82. The northern corridor that Malaysian officials dismiss flies through tightly guarded airspace over India, Pakistan, and even US military installations in Afghanistan, and no one in any of those places has reported a rogue plane. (CNN. com, April 7, 2014)

83. The southern arc under consideration stretches from Indonesia to the southern Indian Ocean and becomes a solid lead for investigators. (CNN. com, April 7, 2014)

84. On March 15, 2014, Malaysian officials announce that the flight veered off course to the west, due to deliberate action.

85. On March 17, 2014, Australian officials take charge of the coordination of the search and

rescue operation, and over the next six weeks, an intensive aerial and surface search is conducted by assets from Australia, Malaysia, China, Japan, Korea, the UK, and the US.

86. During that six-week search period, the Australian Maritime Safety Authority and Australian Transport Safety Bureau jointly determine a search area strategy correlating information from a joint investigation team (JIT2), located in Malaysia, and other government and academic sources.

87. On April 28, 2014, the aerial search is concluded and the search moves to an underwater phase. (Australian report)

88. On March 17, 2014, the initial search area is determined by the joint investigation team to be a 600,000 km² area approximately 2,500 km from Perth, in Western Australia. The initial search area is determined following an analysis of satellite communications data to and from MH370 that was recorded at a ground station in Perth. The data indicated the aircraft flew an additional six hours after its last radar contact with a track south to the Indian Ocean. The area was determined using only limited radar, satellite, and performance data and assumed a southern turn of MH370 at the northwest tip of Sumatra, Indonesia.

89. Areas in the southern Indian Ocean designated S1–S36 were defined from the aircraft's predicted performance and endurance. Two speeds resulted in the longest, straightest tracks to the sixth arc and were used to define possible impact locations within areas S1 and S2. (Australian report)

90. Thai military data reveals the plane had virtually reversed course and took a sharp turn westward, toward the Strait of Malacca. The Thai information bolsters the search in the southern arc, over the Indian Ocean.

91. The Malaysian government says evidence suggests the plane was deliberately flown off-course and traveled back over the Malaysian Peninsula and out into the Indian Ocean.

92. A search of the Strait of Malacca yields nothing. (CNN.com, April 7, 2014)

93. The search in the Australian search-and-rescue zone commenced on March 18, 2014, ten days after the aircraft went missing. (Australian report)

94. A surface search of probable impact areas along the arc, coordinated by the Australian Maritime Safety Authority, was carried out from March 18, 2014–April 28, 2014.

95. The search areas between March and April progresses from an initial southwest location along the arc in a north-easterly direction.

96. The first ocean floor search was completed on May 28, 2014. (Australian report)

97. March 19, 2014: A US official familiar with the investigation says that based on "present search patterns and available data," it is far more likely that the plane would be located in the southern arc. (CNN.com, April 7, 2014)

98. March 24, 2014: The Malaysian prime minister says that evidence made it clear that the flight ended in the southern Indian Ocean (BBC News, March 16, 2014)

99. March 24, 2014: Based on satellite data, the Malaysian prime minister announces in a press conference that flight MH370 ended in a remote area in the southern Indian Ocean.

100. A text message is sent to families thirty minutes before the March 24, 2014 press conference informing relatives of the missing passengers that their loved ones are dead "beyond any reasonable doubt.

101. March 27, 2014: The search for MH370 moves to a different patch of the ocean, 680 miles northeast of the primary area of focus. Australian officials say the move is due to a "new credible lead" provided by Malaysian investigators.

102. The new information was based on an analysis of radar data from the day the plane disappeared. It suggested that the aircraft was traveling faster

than previously estimated before it dropped off radar. The Australian Maritime Safety Authority said this meant that the plane burned fuel faster, shortening its maximum possible distance over the southern Indian Ocean. (CNN.com, April 7, 2014)

103. March 31, 2014: The Australian defense vessel *Ocean Shield* is deployed from Perth. It is equipped with a towed pinger locator system. (Australian report)

104. April 2, 2014: Search vessels with equipment capable of acoustic detections were en route to or near the seventh arc. The areas had been sized so that the primary TPL system embarked on by *Ocean Shield* could cover a particular area prior to the predicted expiry of the flight recorder ULB. (Australian report)

105. April 4, 2014: *Ocean Shield* deploys its first towfish. The towfish exhibits acoustic noise and is required to be changed out with a second towfish.

106. April 5, 2014: The second towfish is deployed and shortly after, while descending, detects an acoustic signal at a frequency of approximately 33 kHz. In all, two ping signals are heard on April 5, 2014. One signal holds for two hours and twenty minutes.

107. April 8. 2014: Two more pings are detected.

108. A ship belonging to the British navy, *HMS Echo*, is tasked to search the area where the pings were received. It finds that the detections were unlikely to have been made by a ULB, due to the depth to the seafloor, surface noise, and equipment utilized. A submarine tasked to the area is unable to get any detections.

109. A review of the *Ocean Shield* acoustic signals was undertaken independently by various specialists. The analyses determined that the signals recorded were not consistent with the nominal performance standards of the Dukane DK100 underwater acoustic beacon. The analyses noted that, while unlikely, the acoustic signals could be consistent with a damaged ULB, and it was decided that an ocean floor sonar search should be performed to fully investigate the detections. (Australian report)

110. April 11, 2014: Australia's prime minister, Tony Abbot, expressed confidence that searches knew the position of Flight 370 within "some kilometers."

111. Based on the analysis of the acoustic detections made by *Ocean Shield*, an underwater sonar survey using an autonomous underwater vehicle commenced on April 14, 2014. It undertook thirty missions, hitting depths from

3,800–5,000 m. The side scan sonar tasking comprised a 10 km radius area around the most promising detection area and a 3 km radius area around the other three detections was completed on May 28, 2014. The total area searched during this time was 860 km^2. It did not lead to the detection of any debris or wreckage. An Australian report said the ATSB considered the search in the vicinity of the *Ocean Shield's* acoustic detections complete and that the area could be discounted as the final resting place of MH370. (Australian report)

112. On April 28, 2014, the aerial search concluded, and the search moved to an underwater phase. (Australian report)

113. April 28, 2014: Tony Abbot announces that the search in the Indian Ocean will be expanded and the operation will employ the help of private contractors. Searchers then face the task of covering about 23,000 sq. miles, about the size of West Virginia. They would switch from autonomous drones to remotely operated submersibles.

114. The ATSB is responsible for defining a search area. Since May 2014, a search strategy group, coordinated by the ATSB, has been working toward defining the most probable position of the aircraft at the time of the last satellite communication.

115. May 1, 2014: The Malaysian government releases documents that reveal that seventeen minutes had passed after Flight 370 disappeared from civilian radar and before ATC in Vietnam and Malaysia raised concerns. (*New York Times* online, May 1, 2014)

116. The Malaysian government confirms that the delay in raising the alert about the plane's disappearance delayed search and rescue efforts, and that they didn't begin until more than nine hours after MH370 disappeared from civilian radar. (*Wall Street Journal* online, May 3, 2014)

117. May 27, 2014: Malaysia's Department of Civil Aviation releases data logs prepared by Inmarsat. In total, forty-seven pages are released but this is not all of the raw data. The first two pages explain how to read the various lines and the forty-five subsequent pages are data that must be interpreted by satellite experts. (CNN)

118. May 27, 2014: Australia and Malaysia release satellite data indicating that MH370 ran out of fuel. The released information, however, is missing metadata that puts this hypotheses into perspective by providing the speed and altitude of the missing aircraft.

119. An Australian report said that MH370's final digital handshake with Inmarsat satellites wasn't at a time that would have made sense for it to go

through, given its previous transmissions. This could be due to the plane's electrical systems resetting if it ran out of fuel. (*Wall Street Journal* online, May 28, 2014)

120. May 28, 2014: The ocean floor search in the planned area around the pings is completed. More than 328 sq. miles have been covered and 97 percent of it has been mapped. (*Wall Street Journal* online, *Joint Agency Coordination Centre (JACC)* statement, May 28, 2014)

121. May 29, 2014: The JACC says that the 850 sq. km section of the Indian Ocean where officials have focused their hunt probably isn't the right place.

122. Australian authorities, who have been coordinating the search in the southern Indian Ocean, say attention will shift to a new area as large as 60,000 sq. km.

123. The underwater search will remain in the same overall region, as officials are basing their focus on an analysis of satellite, radar, and other data that concludes the plane ended up somewhere along an arc stretching into the southern Indian Ocean.

124. Australian officials say they remain confident in the conclusion reached by a team of experts from Boeing and aviation authorities around the world, including the United States and Europe

that Flight 370 ended its journey in the southern Indian Ocean.

125. June 18, 2014: An Australian survey vessel, *Fugro Equator*, joins a Chinese navy ship, *Zhu Kezhen* in mapping the ocean floor using Bathymetric survey. It is anticipated that it will take at least three months to complete the survey of the 60,000 sq. km zone.

Here is a visual of the Accident Algorithm:

THEORETICAL PROBABILITY MATRIX

THEORY VS EVENTS MATRIX

ACKNOWLEDGMENTS

Thanks go out to the following people:

My lovely bride and my son, Tyler, and his family for putting up with me while I wrote this book.

Julia Lord at Julia Lord Literary Management and Judy Coppage at the Coppage Company for their continual encouragement.

Michael Wright at Garson & Wright Public Relations for his support and commitment.

Jeff Zucker, Lucy Spiegel, and rest of the team at CNN.

John Fiorentino for his commitment to the truth.

Michael Exner for sharing his knowledge and balanced insight.

The many Boeing 777 Captains who openly shared their experiences with me.

Salim Ahmad for sharing his exceptional technical knowledge of the Boeing 777.

Trista Wang at Tencent in China for connecting me with 200,000 followers on Weibo who seek the truth about MH370.

Julia Abramoff at Skyhorse Publishing for her hard work and commitment to this project.

Jalal Haidar at World Aviation Forum for providing introductions to the world's aviation leaders.

Dr. Olumuyiwa Benard Aliu, President of ICAO, for allowing an exclusive interview.